TPM
Reloaded

Total Productive Maintenance

Joel Levitt

Industrial Press Inc.

New York

Library of Congress Cataloging-in-Publication Data

Levitt, Joel, 1952-
  TPM Total Productive Maintenance / by Joel Levitt.
    p. cm.
  ISBN 978-0-8311-3426-6 (hardcover)
  1. maintenance—Management. I. Title.
  TS192.L467 2010
  658.2'02—dc22

                    2010000453

Industrial Press, Inc.
989 Avenue of the Americas
New York, NY 10018

Sponsoring Editor: John Carleo
Developmental Editor: Robert Weinstein
Interior Text and Cover Design: Janet Romano

10 9 8 7 6 5 4 3 2 1

Dedication

I want to dedicate this book to the power people have when they wake up, get excited and motivated. The story of TPM is the story of a firm without other advantages than its own people's ideas and motivations, becoming the dominant player in its industry.

Times are tough for manufacturers and doubly tough for manufacturers not fully using their greatest asset. Many companies parrot the idea that their people are their greatest asset, but few act in a way that is consistent with that ideal. That is now the past.

The window of opportunity to change the paradigm of the worker and the manager won't be open indefinitely. It's possible that the greatest enemy is not labor or management but rather the marketplace and the tough realities of manufacturing. Terrible financial woes have made a quantum change not only possible but essential. The companies that wake themselves up and fully use the people at their disposal will be the ones that dominate the next business cycle.

On a more personal note, I dedicate this book to Nancy Boxer. Her love for reading and writing has been a motivator to me. I am lucky she has come into my life. I learn something from her every day.

Thank you, and I hope you enjoy this book

Joel Levitt
March 2010

About the Author

Since 1980 Joel Levitt has been the President of Springfield Resources, maintenance consultants in a wide variety of industries including pharmaceuticals, oil, airports, hospitals, high tech manufacturing, school systems, government, etc. He has worked in both highly regulated (such as FDA, DOT) and unregulated environments, and in both union and non-union environments.

Mr. Levitt is a leading maintenance trainer throughout the United States, Canada, Europe, and Asia. He has trained over 15,000 maintenance professionals from 20 countries in more than 500 classes and workshops, including TPM seminars and projects for the U.S. Mint, MPI (precision stamping), and others. Surveys of his students show that 98 percent rated the training very good or excellent.

Previously, Mr. Levitt was a senior consultant at Computer Cost Control Corp, where he designed and implemented computerized maintenance management systems for organizations such as FedEx, United Airlines, JFK Airport, and BFI. He designed, installed, and serviced a complete automation system with rack control, accounting, and inventory control for BP North America's 30,000-barrel/day-oil terminal. He designed the railroad fuel security and accounting system that was adopted by the American Railroad Association as their recommended standard.

Mr. Levitt has written seven books on maintenance management and contributed chapters to two other books. He has written over six dozen articles for trade publications on maintenance topics.

Table of Contents

Conventions:

While TPM was born in the manufacture of automobiles, elements can be applied to any productive effort. The concepts apply to a bus driver (as the operator), to a control room operator aboard a ship, or in a refinery.

This book is designed for managers and supervisors of operations and production in whatever form. That includes not only traditional manufacturing and assembly, but also mining, refining, smelting, and chemical production as well as service industries such as trucking and facility maintenance. Frequently I use the words operations and production interchangeably. In all cases, I mean both.

TPM Today

I don't want to make anyone angry. But I have some things to say about our productive efforts that might be shocking or rude. This is not personal. The commentaries might not even be true for your organization. If that is the case, please sit back and enjoy the discussion about how others miss opportunities and make mistakes. Of course, maybe a little of the discussion might be true for your organization.

Part of the problem was written almost 500 years ago: "Small problems are difficult to see but easy to fix. However, when you let these problems develop, they are easy to see but very difficult to fix." Machiavelli, Principe (1530)

Another part of the problem is complacency. The techniques we use to produce useful outcomes — such as product production, transportation, or processes — are not perfect. Many developed organically without a master plan or even a road map. Some of them are not even good; the most we can say is that they kind of work. We have to awaken to the challenge (to quote the 14 points of W. E. Deming).

At its very core, TPM shouts wake up! The era of workers dulled into sleeping zombie-like automatons is over. The era of insulated and insular management is over. Now even the most modest line workers have to solve problems, go outside their comfort zones, do maintenance tasks, and work to eliminate waste. Even the most stalwart unions have to break from their own past and embrace the idea that the enemy is in the marketplace, not in the executive suite.

Every manager has to use all the capabilities of all their people to reduce waste, improve reliability and quality, and improve safety. The people to solve problems and be mind workers have been downsized, liquidated or will be soon. What we're left with are the workers, a few managers, and fewer staff positions.

Of course, operators are busy. Everyone in the company is busy. The issue is what is the highest value-added activity they can be doing? In many cases, the highest value-added activity is TPM activity. The same issue can be said for downtime. In some cases, additional downtime has a high return on investment such as when it reduces emergency downtime of an equal or greater duration.

Let's face it: the name TPM confused everyone (me too). What a great idea — let's work hard to make maintenance more productive. Logically everyone wants maintenance that is totally productive. Raise your hand if you want partially productive maintenance!

In a Japanese auto assembly plant, the name makes complete sense. Maintenance is done mostly on the line by contractors (big jobs) or line workers (small jobs). They have only a small maintenance department. In the western world, with actual maintenance departments and sometimes with union rules forbidding operators from using tools to make repairs, the name means something entirely different. If we could go back in time, we might have named it ODI (Operator Driven Improvement), ODR (Operator Driven Reliability), or TYI (Tag, you're it). But we are left with a particularly tough and confusing name.

It might seem funny, but one resistance to TPM comes from the fact it was not invented here (meaning in the West). Some firms resist ideas that are not invented by their company, industry, or country.

TPM has a hidden heart. TPM has a great cover story of moving hours for basic maintenance from operations to maintenance. In our zeal, we have confused the cover of the book with the heart of the book. The cover says TPM moves maintenance tasks to operations. If we go deep inside the book, we realize TPM is less about moving maintenance tasks to operations (as useful as this transfer is) than it is about creating a new accountability for useful output and machine health for operators.

TPM says everyone is involved. Management interprets that as "let's delegate this TPM to maintenance or operations," when, in fact, management is the stick stuck in the ground and will have to change. The change starts by funding and supporting the transition from the workforce as it is to a workforce that is trained (skills, knowledge), empowered, and motivated. It proceeds with managers getting out of the way of the empowered, trained, and motivated workers they helped create. TPM has to overcome the natural conservatism (and laziness) of management. As many shop floor problems (in the trenches) flow from the fear of change as flow from a love for the status quo.

Where we started

For years I've said that TPM does not apply to most North American and European manufacturing facilities. If you follow the original prescriptions as laid out in TPM Development Program and TPM Introduction to TPM by Seiichi Nakajima (1982 in Japanese and 1989 in English, see reference section for details), you will be overwhelmed by the details. That system required charts, displays, detailed flow charts, and an excessive amount of manual record keeping.

The second issue was the overt hostility of TPM to maintenance as a department or profession. In its Japanese incarnation, maintenance is relegated not to the back seat, but to walking behind the car. In a complex, dangerous, mission critical process plant, you want your varsity team doing maintenance — not the intramural team, no matter how great their intentions.

The third issue that was never addressed in the original design was how to interface the TPM activity with the CMMS (computerized maintenance management system). After all, the CMMS is where all the maintenance and repair data is supposed to be kept. Do you issue work orders for the daily TPM work, for the minor repairs, or for the TPM meetings? The Japanese TPM effort took off well before CMMS became so big; therefore, the interface between the two was never discussed in the literature.

The fourth issue was the cultural difference and practical difference between typical Japanese and western companies. In Japanese companies of that era, workers were hired as a class (like the class of 2009). The members of that class periodically moved around to different jobs within the plant. They gained a wide view of their process and had experience both upstream and downstream from where they were assigned this period. The worker was used to being a generalist and quickly adapted to responsibilities of a traditional TPM environment.

In Western plants, people usually stayed in the department in which they were hired. It is uncommon (though certainly not unknown) for people to move between departments. They become specialists in whatever area they ended up. Asking operators to take on even basic maintenance tasks is asking a lot.

Well, what changed?

The economic dislocation of 2008–2009 is what happened. For manufacturing to survive in the West, something had to give. China and the Far East had the advantage of low labor rates and the advantage of seemingly endless supplies of workers. At some point, these workers would demand better wages, but now winning a head-to-head competition was impossible.

In reaction, Western companies were busy cutting costs. Some cut deeply into the muscle of maintenance and other areas. Remember that, in most western businesses, maintenance provides capacity; without the maintenance effort, capacity gradually becomes less reliable, quality varies, and the customer experience suffers.

My colleague at the University of Alabama, Dr. Mark Goldstein, is extremely concerned by this trend. He told me in a recent conversation, "More customers are being lost due to equipment reliability problems than to quality issues. Today, too many companies are losing valued customers because, in their rush to service increasing customer demand, their management overlooked the fact that Just-In-Time delivery depends on full plant equipment availability. Simply, availability is dependent on companies maintaining full plant equipment capacity! Too many senior company executives overlooked their responsibility to strengthen their maintenance knowledge and activity. <u>The result</u>: Customer Loss!"

All products are a child of someone's mind. The processes to produce them are thoughts converted into action. The mind can think of different ways of producing products with different reliances on labor, tooling, and materials. The only way to win the challenge against low labor rates is to outthink the competitor to make labor a smaller part of the equation. To outthink the competitor, you need to harness the intellect of everyone in the company from the CEO to the janitor.

Some people in operations are offended by the following indictment. This indictment is a condition that all humans in all jobs face. One way to engage the minds of the operations group is to wake them up! Most of the shop floor programs for the last 30 years were attempts to wake people up and keep them awake (e.g., TQM, Quality Circles, SPC, Six Sigma).

One wakeup call (first cousin to TPM) is Lean Maintenance. In Lean we attack all the sources of waste. One big source of waste is having maintenance people travel to the assets for short-duration basic maintenance tasks. It is wasteful of their skills as well as intrinsically wasteful when there is already someone there capable of doing the tasks — the operator. After training, operators are ready (possibly not yet willing) and able to do these tasks. Once the operators have been inside the machine, it becomes theirs. For many people, once it is theirs they wake up.

That is the beginning of the shift, but only the beginning. The same shift comes from management. They have to listen and invest real money in the ideas and proposals of the rank and file.

Their listening and investment leads to the second change needed, addressing the enormous waste in production and operations itself. Newly integrated operators could serve as the waste police officers for the whole production effort (I use production in the widest sense). Success requires a radical change in management's view of the workers and, even tougher, a radical change in the way the workers views their own role.

It is this second part of the shift that is exciting. I believe it is also what the Japanese gurus of the last era meant by the program in the first place, but was obscured but the absorption of TPM into maintenance programs by maintenance consultants. Mea culpa on that score.

The advantages of TPM go well beyond the normal interests of maintenance. The concentration on all the losses proves that. TPM belongs in the production/operations bookstore, not the maintenance

bookstore. It is a production/operations effort as much as Lean manufacturing, Total Quality, or any other company-wide production program.

Chapter 1
Introduction to TPM

Who Is TPM For?

TPM is a program for production (or operations in a power plant, for example). It is a manufacturing (or operational) strategy. In a TPM shop, operators are king of the hill. Without operation's full, complete, and unwavering support, evidence of the TPM program will be hard to find a year after installation, or even less. The word *Maintenance* in TPM seems to scare operations people away. If TPM is implemented by or even initiated by the Maintenance department, it will fail.

When we say *operations*, we mean operators, supervisors, production control personnel, managers, and everyone else all the way up to plant managers. The TPM point of view must be the lifeblood of the productive effort and understood by everyone, especially the middle managers. To a large degree, the active support and at least cooperation of the middle management is the most essential element of a successful TPM installation.

The support of production control (production scheduling) is essential because they have to add TPM time into their schedules. Supervisors are essential because initially the TPM tasks have to be assigned, and managed so they are actually done.

Why Is This?

I was training the operators in a large precision stamping factory about TPM. They were excited because they got to finally address issues that had been bothering them, in some cases for years. We made great strides in cleaning the machines and realizing production gains.

The first phase of the training was complete. The team was supposed to continue the TPM activity in the pilot area and gain enough expertise to roll the program out to other areas. But as soon as I left, the production supervisor told the operators, "Well, that was fun, but now it's time to get back to work." In some cases, the supervisors were harder to sell (and more important to sell) than either top management or the operators. When I returned, we had to start over almost from scratch — with the added morale problems — with the formerly excited operations group.

Production Incentives

The last issues concern production incentives. TPM will increase production after the losses are identified and either eliminated or mitigated. During that transition, production might fall for a time. Production incentives have to be adjusted so that performance of the TPM tasks is covered by the incentive program. Otherwise, TPM activity takes bonuses out of people's pockets (in the short term) and will be sabotaged.

So, if you are maintenance professional, this book can help you understand your important role in TPM. But, this book is for your production counterpart.

What Does TPM Do?

TPM (Total Productive Maintenance]) focuses on the barriers to higher production (Exhibit 1-1).

It's simple to describe, but not necessarily simple to do!

We want to get more production at a lower cost out of our existing asset mix by eliminating waste (Lean Maintenance), managing production losses (TPM), and reducing variation in the production process (Total Quality Management). We also want the plant to be safe, nimble, flexible, and a good place to work.

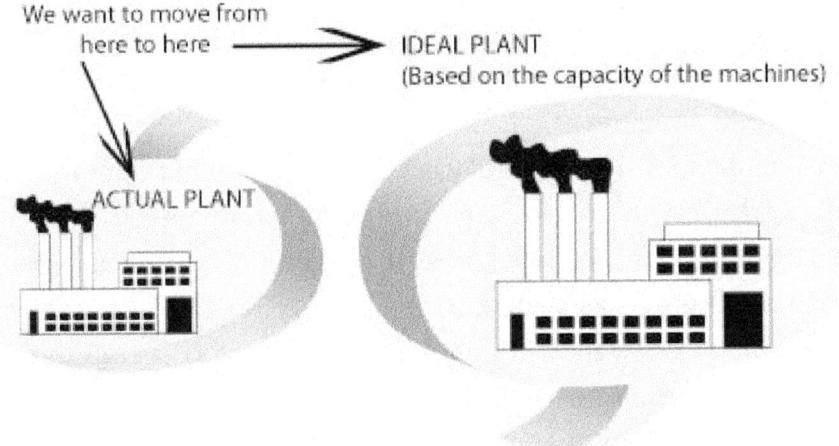

Exhibit 1-1 TPM is Very Simple

Why the Results from TPM Are Urgent Today

A new situation has developed in the way we look at organizations. Throughout the last 25 years, organizations have slashed their ranks, reduced overheads, and optimized processes. This process accelerated at the end of the 2000s with the economic crisis. At the same time, we increased the complexity and speed of the equipment and our reliance on computers, programmable logic controllers (PLCs), and sophisticated controllers. We are faced today with smaller crew sizes — and basic maintenance demands are going unmet. Yet basic maintenance continues to be essential for reliable performance.

Superficially, TPM recruits operators into the maintenance function to handle basic maintenance tasks and become both the champions of their machine's health and the tinkerers to reduce production losses from all sources. Operators become involved in maintenance activity (as well as other areas to detect and mitigate the losses) and decisions. TPM uses the operators in autonomous groups to perform all the routine maintenance, including cleaning, bolting, routine adjustments, lubrication, taking readings, start-up/shut down, and other periodic activities.

This might sound ungrateful (based on the fact that operations personnel work very hard and are proud of what they have accomplished), but at a deeper level TPM is designed to wake up operators and their supervisors as well as everyone up the chain of command. Many people sleep walk through their day and stop seeing the waste right in from of them. That is just not good enough anymore. TPM along with Lean Maintenance, TQM, and RCA (Root Cause Analysis) are complimentary programs that try to get full engagement from the operators and supervisors. With full engagement, success is possible.

The reason for this sleep walking is psychological. The single biggest barrier to identifying waste is called "the normalization of deviance." This is a fancy way of saying that if you look at a pile of junk long enough, you'll stop noticing it. The pile begins to seem normal. This tendency is the bane of operators (who stop seeing minor jam-ups and other missed opportunities), PM inspection (when people stop seeing deterioration), and lean maintenance efforts (when people stop recognizing waste).

Sometimes we have to shake up our normal methods of seeing. TPM is first an extended class in seeing and investigating our productive effort, and second a continuing alarm clock to awaken the producers. But make quick productive use of your altered sight — in a day or two, most you slip back to being blind to the waste.

The maintenance department becomes an advisory group to help with training, setting standards, doing major repairs, troubleshooting, and consulting on maintenance improvement ideas. Maintenance departments are the specialists in major maintenance, major problems, problems that span several work areas, and trainers. Under TPM, maintenance becomes very closely aligned with production. For TPM to work, maintenance knowledge must be disseminated throughout the production hierarchy.

The old philosophy of "Produce at all costs, damn the torpedoes — full speed ahead!" will fall flat on its face with TPM. TPM needs some downtime in order to be successful. The operators must have complete, top level support throughout all phases of the transition and thereafter.

Good maintenance practices, as highlighted by the article in Exhibit 1-2 on the following page, can contribute to the productive output and profit of the whole organization. It contributes to safety too.

Revolution

TPM is revolutionary. It is a game changer on factory floors of organizations that can go all the way to autonomous maintenance. The ideas of TPM are to make the operator a senior partner in the produc-

SAN ANTONIO, March 25, 2009 (Reuters) - A process safety program has improved the reliability of Chevron Corp's (CVX.N) 2.2-million barrel per day (bpd) refining system and yielded a gain in throughput equivalent to a new refinery, Jeet Bindra, Chevron's president of global manufacturing said on Wednesday.

"We found if you operate your refinery reliably and safely, you can produce as much as one new greenfield refinery," Bindra said in a speech to the Hart World Refining & Fuels Conference.

Chevron has reduced unplanned operational interruptions by 40 percent in the past two years through regular planned maintenance and preventive maintenance, he said.

Small increases in operational reliability at each refinery from the safety program have equaled a capacity gain between 100,000 and 150,000 bpd, Bindra said.

"Now is not the time to cut corners," he said. "You need to know the condition of your equipment."

Earlier this year, the U.S. Chemical Safety Board called on refiners and chemical plant operators to not reduce maintenance as they cut costs to counter falling refining margins.

Cost-cutting in equipment maintenance was seen as a factor in the deadly March 23, 2005 blast at BP Plc's Texas City, Texas, refinery, the CSB found.

Chevron is also looking at increased operational automation of operations to improve reliability at its refineries, he said.

"We think the technology is available today to react to upset conditions without human interaction," Bindra said.

Operational control has been centralized in the past two decades in refinery command centers with operators monitoring computer screens and reacting to malfunction alarms with gauges and control valves on refinery units as back-ups. (Reporting by Erwin Seba; Editing by Marguerita Choy)

Exhibit 1-2

tion effort. These ideas, imported from Japan, have taken root in factories, refineries, mills, and power plants throughout North America. They succeed because they force us to realize we have to use more and more of the capabilities of every employee (and vendors too!) to remain competitive. Operators are traditionally viewed as underutilized in most factories.

The machine operator is the key player in a TPM environment. Many of the losses are under the control of the operator, involve the operator, or happen while the operator is near the machine. There is less reliance on the maintenance department for basic maintenance (but more for maintenance prevention projects, productivity improvement projects, coaching, training, problem solving, and mentoring). Control and responsibility are passed to the operators.

Although the operator is the key player, it's the management that ultimately has to make the right choices. Keith Rimmer, a consultant from the global consultancy Woodhouse Partnership, says convincingly that for a company to be successful in Asset Management, "It requires processes that are driven effectively by top management and supported by empowered and competent employees. A key characteristic of successful asset management is consistently making sound decisions and good compromises, and carrying out the appropriate tasks at the right time and at the optimum level of expenditure. Above all it requires the commitment of top management, and it is unlikely that an organization will successfully integrate and optimize their asset management without such commitment."

One interesting side effect was the application of TPM principles to complex process environments such as power plants, sewer treatment and water plants, and chemical plants. The operator of a sewer treatment plant is a very different person than the operator of a machine. The sewer plant operator has studied for and obtained a license. These positions generally require personnel with higher qualifications than do machine operator positions. Some of the same ideas can be applied, but the tasking has to be reviewed by personnel knowledgeable about both plant operation and maintenance. In a later section we will review the reason TPM was developed in automobile assembly plants and not in other business types.

Legitimate Fear

In today's cut-happy environment, any change is viewed through the lens of job loss. TPM is no exception. When Paul Wilson, Managing Director of Aster Training, installed a TPM system, his team had to face fear from the maintenance ranks. He said, "We worked hard at alleviating any fear the maintenance technicians might have had that their jobs were under threat. This soon evaporated once we were challenging them with more interesting projects." Much of the work being taken over by operations is not being done at all (like basic maintenance activity). Maintenance departments will find themselves with ample work on just the projects requested by the TPM teams. These projects will generally reduce waste, make the machinery operate better, or achieve lower levels of product variation. The result is better profit without personnel cuts (for the same volume). Of course significant cuts in volume will require retrenchment, but that is another conversation and should not be mixed up with a TPM implementation.

History of TPM

One of the leading thinkers in TPM is J. Venkatesh. His review of the history of TPM can be found in an article on Reliabilityweb.com.

TPM is an innovative Japanese concept. The origin of TPM can be traced back to 1951 when preventive maintenance was introduced in Japan.

W.E Deming traveled to Japan under the Marshall Plan and began a powerful and eventually world-changing discussion about quality. The royalty of what was left of Japanese industry were in the training rooms. They saw that the drive for quality and efficient production was their only edge. They had no resources beyond what they could build and imagine.

The ideas of TQM (Total Quality Management) and the concepts of PM were only partial answers to the issues of maintenance and quality. After much trial and error, the PM effort evolved into TPM.

The concept of preventive maintenance was taken from American industrial practices. Nippondenso* of the Toyota group was the first company to introduce plant-wide preventive maintenance in 1960. Preventive maintenance is the concept where operators produced goods using machines, while the maintenance group was dedicated to maintaining those machines. However, with the automation of Nippondenso, maintenance became a problem as more maintenance personnel were required. So the management decided that the routine maintenance of equipment would be carried out by the operators. (This is the beginning of Autonomous Maintenance, one of the features of TPM). The maintenance group took up only the essential maintenance works.

---

* Manufacturer of auto parts and part of the Toyota group of companies. Was established December 16, 1949 as Nippondenso Co., and subsequently renamed DENSO. As of March 31, 2005, DENSO Corporation consisted of 171 subsidiaries (64 in Japan, 33 in the Americas, 31 in Europe, and 43 in Asia/Oceania) with a total of 104,183 employees. In 2006, DENSO was listed at #207 on the Fortune 500 list.

Thus, Nippondenso, which already followed preventive maintenance, also added autonomous maintenance done by production
operators. The maintenance crew was released of their routine maintenance tasks; instead, they carried out equipment modification for
improving reliability and maintainability. The modifications were
made or incorporated in all their new equipment. These tasks are
aimed at maintenance prevention (MP). MP or Maintenance
Prevention is the elimination or reduction in the need for maintenance. By reducing the source of the dirt, we also reduce the need for
cleaning (an example of Maintenance Prevention). Thus, preventive
maintenance along with maintenance prevention and Maintainability
Improvement gave birth to Productive Maintenance (PM). The aim of
productive maintenance was to maximize plant and equipment effectiveness to achieve optimum life cycle cost of production equipment.

By then, Nippondenso was using quality circles, involving the
employee's participation. All employees took part in implementing productive maintenance. Based on these developments, Nippondenso was
awarded the Distinguished Plant prize by the Japanese Institute of Plant
Engineers (JIPE) for developing and implementing TPM. Nippondenso
became the first company to obtain the TPM certification.

TPM had become a part of TPS (Toyota Production System).
According to Bob Williamson, a long-time veteran of the TPM wars,
the TPS systematically focuses on the identification and elimination of
waste to reduce manufacturing costs. In Japanese plants of that era, the
culture was ripe for involvement of everyone in the production
process.

In many ways TPM is a return to a pre-1920s model of maintenance. Before the 1920s, machine operators were skilled mechanics,
so they were expected to repair their own machines. As mass produc-

tion took over, lower-skilled operators were recruited and the production jobs became more menial. Many of these newly-minted operators were immigrants or just off the farm. Their greatest advantage was their lower wage rate and the long hours they were willing to work.

As the numbers of machine operators grew, the ability to fix one's own machine quickly was gone. Company-sponsored training to improve one's skills was non-existent. Soon this group, as well as management, forgot that these people had capabilities far exceeding those needed as operators. A tradition settled in of operators being only button pushers. The maintenance department as we know it developed at that early time by necessity, filling in with specialists in repairs and maintenance.

A new situation has developed in the way we look at organizations. For the last 35 years, organizations have slimmed ranks, reduced overhead, and optimized processes. At the same time we have increased the complexity and speed of equipment and our reliance on computers, PLCs, and sophisticated controllers. We are faced today with smaller crew sizes and greater maintenance demands than ever before.

TPM recruits the operators into the maintenance function to handle basic maintenance tasks and to become the champions of the machine's health. TPM returns to the pre-1920 roots by re-involving the operator in maintenance activity and decisions.

The maintenance department becomes an advisory group to help with training, setting standards, doing major repairs, and consulting on maintenance improvement ideas. Under TPM, maintenance becomes more closely aligned with production. For TPM to work, maintenance knowledge must become disseminated throughout the production hierarchy. The operators must have complete, top-level support throughout all phases of the transition and thereafter.

TPS,(Toyota Production Systems) the Parent of TPM

TPM was the brainchild of Toyota. It was based on a component of the Toyota Production System. The production system had several parts. TPS (as it is called) is designed for a particular time, circumstance, and place. It was designed by people in tune with their culture and organization. Finally, TPS was specifically designed to manage the assembly of automobiles.

Machines to be Worked Upon

There are special attributes of automobile assembly that lend themselves to a TPM approach. The first item is that the tools are pretty small (compared to other industries such as steel making or mining). These relatively small tools meant that the operators could literally put their hands on the machine and learn something about its operation.

The second factor was that these tools and the processes used are not intrinsically hazardous (with a few expectations). This means that you might get hurt if you do the wrong thing, but it is unlikely that you will endanger others, like you could in an oil refinery or mine.

With the assembly tools being smaller, the need for elaborate sets of repair and maintenance tools and deep maintenance knowledge is not as important, especially for basic maintenance services. Finally, the machines are modular and quite sophisticated. When something breaks, usually an entire module is replaced.

The Employees

There are two aspects of auto assembly employees. This group is usually well paid and has low turnover. Low turnover is essential for TPM because of the training required. High pay means of the people attracted to factory work, the top tier is attracted to the automakers.

The Business System

In Japan most heavy, hazardous, and complex maintenance is done by contractors. The individual plants usually don't have a deep or large maintenance department, except for a few mission critical services.

Of course, one big reason TPS thrived was that it was supported from the top of the organization. TPS permeated all activities of Toyota. It was patiently nurtured, revised, and improved until it entered the company's DNA and it was expressed in most decisions made in the plant.

For these reasons, TPS was a logical system that was adopted and supported over a long period of time by Toyota's management.

TPM, TQM (total quality management), Lean manufacturing, Lean maintenance, and JIT form the basis of the Toyota Production System (TPS). All of the programs dovetail together and support each other. In the end, the company produces high quality products with as few inputs as possible and as little waste as possible. For more information on this fascinating story, consult *The Machine that Changed the World* (bibliography).

The TPS House Graphic

"One of the most recognizable symbols in modern manufacturing is the 'TPS House' diagram as shown below. The diagram (Exhibit 1-3) is a simple representation of the Toyota Production System (TPS) that Toyota developed to teach their supply base the principles of the TPS.

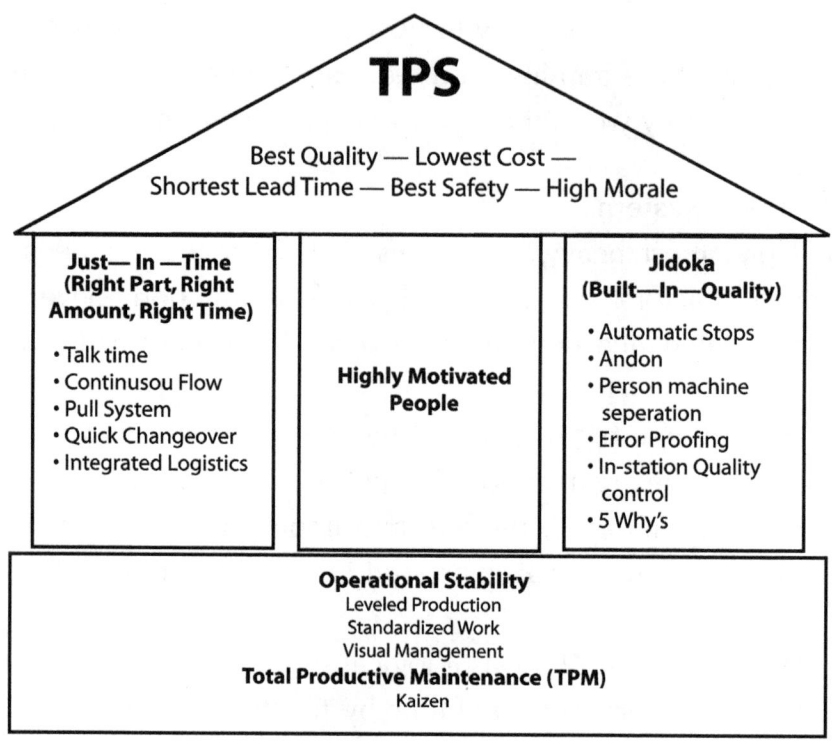

Exhibit 1-3 The TPM House

The foundation of the house represents operational stability and has several components, one of which is Total Productive Maintenance." From an article by David McBride at the Reliable Plant web site (for the complete article, go to: http://www.reliableplant.com/Article .aspx?articleid=8417).

The house graphic shows the relationship between all the parts that make up TPS. It also shows TPM to be a foundation activity necessary for success. The results are the roof which when achieved becomes a competitor killer!

The Toyota system is designed to remove waste from the production of automobiles. It is interesting that the Japanese words become like Zen koans (stories) that disciples study to understand the mysteries. In this case, the Japanese words identify a type of waste. For example one type of waste is overburden and the Japanese word is muri. The waste associated with doing things differently each time (inconsistency) is called Mura.

All types of waste are called Muda. The challenge is designing a process capable of delivering the required results smoothly; by designing out "mura" (inconsistency). The design ensures that the process is as flexible as necessary without having to overproduce "muri" (too much work-in-process or overburden) since this generates "muda" (waste). There are seven kinds of muda that are addressed in the TPS:

1. overproduction
2. motion (of operator or machine)
3. waiting (of operator or machine)
4. conveyance
5. processing itself
6. inventory (raw material)
7. correction (rework and scrap)

The elimination of muda has come to dominate the thinking of many when they look at the effects of the TPS because it is the most familiar of the three to implement.

The entire organization is aligned to solve the most pressing problems that get in the way of high quality/ high quantity output. TPS has several sub-programs (of which TPM is one).

JIT

JIT is an advanced method of regulating production. Using JIT, only a small number of parts are made at one time. In some cases, it might be enough parts for 4 or 8 hours of assembly line production. These parts are replenished on a just-in-time basis. That means when the last part is assembled onto the product, a pallet of the next batch of parts is put down. The overriding issue is that when the machine is needed, it had better work or the line will be shutdown in a short time (like 4 or 8 hours). High reliability and quick repair are essential elements of a JIT environment.

When Harley Davidson, the leading American motorcycle maker, became an independent company, its management decided that JIT manufacturing was the best solution to the quality problems they were experiencing at the time. The thought was with minimal part runs a quality fix would get onto motorcycles within days. They used to run months of parts at a time to optimize and amortize the set-up time. Under the old plan, it might be months before the parts and sub-assemblies in stock were used up and the improved part appeared in bikes.

Motorcycles were assembled in York, PA, and the engines and transmissions were made and assembled in Milwaukee, WI. The truck of engine and transmission assemblies was supposed to arrive when the York plant got down to a shift of stock. In other words, if the truck delivering the engines was more than 8 hours late, the York assembly line would have to shut down.

Now imagine running JIT with equipment that was only 70% reliable. Three times out of 10 the equipment would be down when you needed it to produce parts.

"Total productive maintenance (TPM) is indispensable to sustain just-in-time operations," says Dr. Tokutaro Suzuki, Senior Executive Vice President of the Japan Institute of Plant Maintenance in *TPM in Process Industries*. In a JIT system, he emphasizes, "You have to have trouble-free equipment." Prior to the adoption of TPM, Japanese manufacturers found it necessary to carry extra work in progress (WIP) inventory "so that the entire line didn't have to stop whenever equipment trouble occurred. The concept is that the operator must protect his own equipment," he explains. "Thus the operator must acquire maintenance skills."

However, maintenance experts may still make periodic inspections and handle major repairs. Design engineers also play a big role. They must take maintenance requirements — and the cost of equipment failure — into consideration when they design the equipment, stresses Dr. Suzuki.

Another Japanese Vocabulary Lesson

Using the Japanese language, TPM focuses the energy of the organization on the actual piece of work (gemba), the actual part (genbutsu) being made, and the actual activity performed or as the Japanese translation says phenomenon (gensho).

TPM directly attacks the sources of ineffectiveness by concentrating on all the sources of loss of production (not only on the maintenance-related losses). This is important because most mistaken impressions think the focus of TPM is on maintenance whereas the true focus is on high quality output.

Where Does Efficiency Fit In?

Most companies spend enormous amounts of money on improvements in efficiency. Efficiency is defined as doing things the right way.

Industrial engineers spend a great deal of effort on insuring wasted movements are eliminated from the production process. TPM can be said to take the next step. TPM looks at doing the right things right. By attacking all of the losses, TPM insures that at the end of the day, the pile of saleable parts made by the process is bigger. In some cases, after TPM, the pile of good parts is a great deal bigger.

Some Important Questions

Where did the idea for TPM come from in your plant? As mentioned, if the maintenance department initiated the program, look out. It is difficult to convince anyone that the program is not just a way to off-load maintenance work onto operations. Related to this question is another: Who is driving the effort?

What difference does your situation make to the success of TPM? If your machines are large or hazardous, or they require enormous skill to even consider fixing, then traditional approaches to TPM will not easily work. The fundamental shift is not maintenance activity to operations (which everyone seems to focus upon) but the responsibility for all production losses shifts to operations.

How disciplined is your production effort?

Although not every company needs full TPM implementation, every company has something to learn from it.

Chapter 2
Talking About TPM

Total Plant Involvement

Total

Before we can really talk about TPM, we have to define and dissect the words. The word total means we, as an organization, are all in. If this was referring to a sport, it would suggest we are going to play full out. One of the dictionary.com definitions says, "involving all aspects, elements, participants, resources, etc.; unqualified; all-out: total war." We are in total war against deterioration and anything else that detracts from the output.

Another definition is "constituting or comprising the whole; entire; whole: the total expenditure. Of or pertaining to the whole of something...." This means that the whole organization has to align itself to support the TPM process, products, and activities.

The last definition is how deep do we go? The answer is "complete in extent or degree; absolute; unqualified." We are all in to make TPM work.

Productive

   If we are productive, we "have the power of producing; generative; creative: a productive effort." The TPM effort is producing more output with the same or fewer inputs. So we are generating something. The second definition is important for our understanding. Productive is "causing; bringing about." This effort will bring about some important changes.

Maintenance

   Finally we have the word *maintenance*. Maintenance activities mean activities that are designed to keep an asset in good condition and not let deteriorate in the first place. If we maintain our weight, it means that we kept our weight the same. Many people think that maintenance is a fancy word for repair or fix. Nothing could be further from the truth — maintenance is activity that avoids the need to repair. If repair is necessary, then maintenance has not been present.

   Because the scope of TPM is well beyond the concept of maintenance, we might want to call the effort something more accurate like Total Output Management. But I can leave the naming of your program up to you.

   Although TPM, in its entirety, doesn't apply in many situations, aspects do apply to all maintenance situations.

A Day in the Life of a TPM Shop

   At the U.S. Mint, Jim Dunn (a composite person, not a real operator) walks into the coining shop, where he has worked for the last 11 years, to start his shift. At the beginning of every shift he takes a few minutes, starting at one end of the line of his presses, to clean, check the lubricant level and listen to each press. The routine is designed so that there is minimum interruption of the press. If a tool change is imminent, he will do his inspection during that period. He will also time his service to coincide with the end of each batch of material.

He can hear the pitch of the high-speed presses. From his experience he knows that press #1 is running slow and 2, 3, 4 and 5 are right on the money (so to speak). He verifies his hunch by looking at the LCD readout on the Press Control Center. Presses are slowed down sometimes to reduce the number of problems.

Jim knows from his TPM training that "running slower than specification" is one of the losses that it is his job to track down. Making sure all the presses are running correctly, he calls the TPM coordinator for the coining department. They discuss the slowdown and decide to run the press at full speed and watch it closely.

After three hours of perfect operation, the blank feeder jams up. Jim calls the TPM coordinator and they agree to look into the problem. The problem is cleared and the machine is turned back down until the next day, when the team can meet and take a look. Jim calls the floor supervisor and tells him that Press #1 will be out of service the next day for a few hours.

Jim comes in the next day and goes through TLC (Tender Loving Care which translates to Tighten, Lubricate, Clean) on his other presses. A relief operator is assigned to the other presses while the TPM team looks at Press #1. Step one is a thorough hands-on cleaning and inspection. The feeder is looked at very closely.

It becomes clear that the feed fingers have been messed up and are slightly bent. The feeder tube also has some irregularities, and five bolts are loose. These problems are cleared up; then the machine is run up to full speed. The press is stable for the remainder of the shift. Jim tells the second shift operator to keep an eye on Press #1, and reminds her to clean up coffee cups left inside the sound enclosures.

Jim goes home feeling that he made a little, but real, contribution to the organization that day.

Promises Made at a Recent TPM Conference

As discussed previously, oil refining is not a natural home for TPM. Nevertheless, the adoption of certain precepts will make significant improvement possible. The following precepts were developed recently at a TPM conference for oil refiners in the Persian Gulf:

- Manufacturing equipment uptime: up 40%
- Unexpected equipment breakdowns: down 99%
- Equipment speed: up 10%
- Defects caused by equipment: down 90%
- Equipment output (productivity): up 50%
- Maintenance costs: down 30%
- Return on investment: increased several hundred %
- Safety: approaching zero accidents
- Improved job satisfaction

Other TPM targets included:
- Obtain minimum 80% OPE (Overall Plant Effectiveness).
- Obtain minimum 90% OEE (Overall Equipment Effectiveness).
- Run the machines even during lunch. (Lunch is for operators and not for machines.)
- Operate in a manner so that there are no customer complaints.

- Reduce the manufacturing cost by 30%.
- Achieve 100% success in delivering the goods as required by the customer.
- Maintain an accident-free environment.
- Increase the suggestions from operators by 3 times*.
- Develop multi-skilled and flexible workers.

Improvement in the Delivery of Maintenance Service

For one second, let's examine TPM and delivery of maintenance service. It is argued by TPM professionals that much of the pure labor benefits flow from the simple fact that the operators are already in front of the machine with the tools and materials whereas the maintenance personnel have to travel. That is true, but not the whole story. TPM is one of the most effective methods of improving the delivery of maintenance service, largely eliminating the time needed for custody transfer, job instructions, travel, and collecting tools and materials, while increasing the effectiveness of the equipment.

In fact, a simple PM Service can take 3 times longer for maintenance personnel than for operations personnel (even at the same depth).

Example:

When maintenance workers do a PM on a piece of equipment, they must go through a variety of steps. These steps are essential, but not productive. In this case, a 45-minute PM might actually take over 400 minutes. In a TPM environment, the same job might be done in

* What goes along with this is to implement as many of the ideas as possible

1/3 of the time (Exhibit 2-1).

If this time savings were the only benefit, the program would be a good solid single or double. With the other benefits in improving the OEE (described in full in Chapter 6), TPM is a home run.

Comparing PM by Maintenance and the Same PM by a TPM Team Member

Activity	Time for a 30-min PM by Maintenance	Time for a 30-minute PM by Operator under TPM
Get job assignment	10 minutes	5 (or none)
Obtain Permit	60	0 (no permit because custody is not transferred)
Lock out tag out	45 (operations and maintenance lock out)	30 (operator needs to be the only one to lock)
Collect tools	15 (special tools in tool crib)	5 (specific tools right there)
Collect parts, materials	15	5 (any materials nearby)
Travel	25 (must travel and return)	0 (already there)
Quick safety walk down	10 (must look at everything)	5 (already works there and knows what has changed)
Perform work	45	45
Clean up work area	10	10
Release permit, Unlock, return to operations	15	5 (no permit, 1 lock)
Do paperwork, work order, closing comments	10 (full PM ticket)	5 (limited TPM ticket)
	400 minutes	115 minutes

Exhibit 2-1 Comparing PM by Maintenance and the Same PM by a TPM Team Member

Selling TPM to Different Stakeholders

If we want to sell TPM within a company, we have to discuss the outcomes that would attract different stakeholder groups. Each such group has different interests, responsibilities, and concerns. We also have to directly address the fears that this kind of change brings on.

Top management and shareholders

The first stakeholder group is management (top management). This group is concerned about profit, getting product out the door, long term viability, and safety. They are sensitive to public opinion and the opinion of shareholders. They also look at any program as a cost that had better provide a return on investment. Last, this group might have bonuses tied to profitable output. Therefore, they will not want to rock the boat unless they see a significant advantage for themselves.

The essential question is how would we prove these benefits? One of the cores of TPM is a rigorous approach to the measurement of all the production losses. For example, the first benefit below (reduced breakdowns and emergencies) is a metric that can be generated from the CMMS for the pilot area. For the other benefits, the metrics are available in TPM, CMMS, or other systems as shown.

Benefits of TPM for top management and shareholders sticking to TPM for the long haul

- Reduced equipment breakdowns and emergencies: CMMS
- Improved equipment effectiveness and throughput: OEE
- Improved product quality: OEE or quality system
- Improved safety: Safety metrics
- More emphasis on preventive work: CMMS PM Performance
- More emphasis on getting to the root of the problem and eliminating it: CMMS looking for a reduction of repeat repairs

- Lower operating costs: Look at utilities used per unit of product produced
- Reduced raw materials, in process, and finished goods inventories Various inventory levels
- Reduced wait time for maintenance craftsmen: Work Sampling study
- Significant improvement in equipment availability and capacity: OEE
- Improved equipment lifespan: Accounting records
- Improved plant productivity: Increased profit from plant
- Improved decision-making involving employees: Survey supervisor
- Higher morale from improved job satisfaction and : job security Survey employee
- Greater ability to meet customer's quality and delivery needs: Survey customers and sales

Operators

The second group that has to be sold on the project is the operator group. Superficially, TPM looks like extra work. Their first reaction might be that "they are trying to get more work from us without more pay." In fact the work is different, rather than more. Operators are still there for their whole shift. The only difference is that they will be taking a larger role in their job and in the success of their company. Ultimately, TPM is more satisfying for the operator.

Benefits of TPM for Operators
- Increased decision-making regarding equipment and processes
- Operators become more valuable to company
- Higher morale from improved job satisfaction and job security
- Operators receive additional training
- More cooperative work environment from team work experience
- Higher level of expertise in production
- Increased self esteem from better performance

- Multi-skilled operators in higher demand
- Ability to address issues that plagued operations
- Better equipment availability and reliability
- Better relationships with maintenance
- More appreciation from maintenance of complexity of the production job

At a recent installation for TPM, Paul Wilson, the Managing Director of Aster Training, described the real benefits for the operators.

- Operators no longer had to wait for maintenance technicians to fix trivial problems. The maintenance guys were no longer required to complete tasks that they considered mundane.

- Maintenance technicians, relieved to be rid of their mundane tasks, are given specific projects working with the production staff to cut out some of the biggest loss areas.

- The machine set-up routines were looked at; new purpose-made jigs were constructed to allow production to carry out changeovers without the need for all of the "tweaks" and alignment problems that had previously plagued this process.

- The major parts of the machines prone to wear and tear were put onto a predictive maintenance schedule and their spares managed on a statistical inventory control basis. In fact, one company was even able to sell back some spares to the machine manufacturers which had been previously been bought as insurance.

- The cleanliness and reliability of machines improved significantly as operators took on a series of daily, weekly, and monthly fitness checks. We encouraged the operators to feel a sense of ownership toward their machines and to treat them as their own, something that would have been an alien concept only six months before.

Supervisors and Managers

Related to the operators are their supervisors and managers. This stakeholder group may be significantly harder to convince about TPM. The key to this group is to be sure to incentivize TPM achievements. If the old incentives are kept in place, then the middle managers will be convinced that top management really wants the old outputs. Without that change, TPM looks like icing and not the cake itself.

Benefits of TPM for Operations Supervisorsand Managers

- Higher-skilled operators
- Smoother production
- Easier to meet goals when everyone is pulling the same direction
- Higher worker morale from improved job satisfaction and job security
- More cooperative work environment from team work
- Better relationships with maintenance
- More challenge and more fun

Maintenance Department

The maintenance department can be a great asset in TPM or an anchor to old ways of working. They are now team members lending specific expertise to the TPM team. They are relieved from having to do basic (boring) maintenance and PM. They can concentrate on higher level issues. We also have to address the fears that this kind of change brings on in maintenance workers.

The obvious linkage is between the transfer of workload and near-term layoffs of expensive maintenance workers. This is a real fear that must be addressed. If there are going to be layoffs, please hold off on your TPM adoption until the labor situation settles down. In fact, the workload might increase after TPM is even partially operational because maintenance will be working on projects geared toward improvements of various kinds.

Benefits of TPM for Maintenance Personnel
- Crafts people receive additional training
- More time available for high level Preventive and Predictive Maintenance, projects, and analytics.
- Higher level of expertise in maintenance
- Better relationships with production people
- Crafts people become more valuable to the company
- More cooperative work environment from teamwork
- More fun because of new relationships and sharing problem solving
- Less routine work for lubrication, cleaning, adjustments, inspection, and minor repairs
- Greater ability to troubleshoot and given time get to the root cause
- More appreciation from production of complexity of maintenance job
- Higher morale from improved job satisfaction and job security

The Value Calculator for TPM and Lean Manufacturing

Exhibit 2-2 on the following page shows a sample of the form for estimating costs, benefits, and ROI. This form can be found at http://wcm.nu/economy/tpm.htm

To estimate the return on investment of your TPM project, first cal-culate the current OEE (Overall Equipment Effectiveness). OEE is the

Return on Investment Calculator for TPM and related programs						
	Today	Year 1	Year 2	Year 3	Year 4	Year 5
Planned Actions		5s	TPM	Kaizen	SMED	Kanban
Estimated OEE Progress	50	55	65	75	80	85
Total Sales	100					
Costs of Raw Materials and Energy	30					
Cost of Production Personne	I30					
Maintenance Costs	10					

Exhibit 2-2 Return on Investment Calculator for TPM and related programs

ratio of the actual production divided by the ideal production for a machine. OEE is described in detail in chapter 6.

Next, estimate how OEE will increase over the years. Then exchange the default values in this form. All figures are in millions of dollars.

You can use the form from Exhibit 2-2 to estimate the value of the concept you are looking at. The main costs will consist of:

- Training and consultancy
- Increased initial maintenance costs
- Project team members' time
- Support activity (IT, accounting, and engineering)

Benefits

A change project such as TPM must be looked upon as an investment that, despite initial costs, will bring something back in return. In this regard, an investment in a change project is no different than any other investment. The question is what to do with the increased OEE.

- Increased production (if your market allows).
- Reduce total number of employees in production (One rule is that layoffs will not happen from these improvements.)
- Some combination of above

Increased OEE ratio is the main factor that that may be used to approximate the return on the efforts. The OEE ratio is a direct reflection of your plant's capacity. Therefore, it may be used to calculate future productivity after completing improvements.

Suppose a plant produces 10,000 units per year with an OEE ratio of 50 percent. After the improvements, the project team estimates that it will be possible to reach an OEE ratio of 80 percent. This means that they will be capable to produce

$$10,000 * 80/50 = 16000 \text{ units in the same facility}$$
and with the same labor as before

Is There a Market for Expansion?

One important question asks whether there is a market for expansion. If so, then the increased capacity may be used for increased sales. It is common that the company's market share might grow after implementing TPM or Lean Manufacturing. This is possible as improved delivery accuracy and shorter lead times make more sales possible, even if the market is stagnant.

If expansion is not considered possible, the increased capacity may instead be used to lower the production costs. This is possible through:
- less overtime
- fewer shifts
- fewer parallel production lines maintained and operated

The direct labor costs for production will, therefore, decrease as the OEE ratio increases. For the example described at the beginning of this section, Exhibit 2-3 shows the result of the calculations.

Increased OEE Ratio at Company; Return On Investment (Millions of dollars)

	Today	Year 1	Year 2	Year 3	Year 4	Year 5
Planned action		5S	TPM	Kaizen	SMED	Kanban
OEE	50	55	65	75	80	85
Total sales (Millions $s)	100	110	130	150	160	170
Costs of raw material and energy (variable)	-30	-33	-39	-45	-48	-51
Labor costs (production)	-30	-30	-30	-30	-30	-30
Maintenance costs -	10	-12	-11	-9	-8	-7
Cost of project team members	0	-0.1	-0.1	-0.1	-0.1	-0.1
Cost of consultancy and training	0	-0.1	-0.1	-0.1	-0.1	-0.1
Incremental Cash Flow	0	4.8	19.8	35.8	43.8	51.8
Cumulative Incremental Cash Flow	0	4.8	24.6	60.4	104.2	156

Exhibit 2-3

The estimation above assumes that the productivity increase can be used to increase production and sales. The payback period is less than 1 year.

Chapter 3
Elements of TPM

TPM Has Two Aspects

1. A rigorous approach to achieving high machine utilization and accurate measurement
2. A shop floor philosophy based on encouraging operators to take a greater role in the health of their equipment and the productivity of the manufacturing process

Autonomous Maintenance Concept (Jishu Hozen)

Maintenance is entirely driven from the TPM team. Enemies of TPM construe autonomous maintenance to mean they can dump all the nasty work no one wants to do onto the operators. They feel they can now send a list of instructions to be carried out. Actually, autonomous maintenance means that the team makes up its own mind about what care is needed for maximum uptime and quality output. This means that management must trust the judgment of the team and the team must be trained to some level of maintenance sophistication.

Maximize Overall Equipment Effectiveness

TPM has a very strict definition of effectiveness called OEE (Overall Equipment Effectiveness). One of the tenants of TPM is that sloppy or non-existent metrics can cover up opportunities for production improvements.

Establish a Shared System of PM

This system should cover the equipment's complete life, taking into account the age and condition (called life cycle) of the equipment. PM should be modifiable based on the age, life stage, and useful life of the equipment. Without this, the PM tasks might not reflect the failure modes of equipment in that condition.

The shared PM divides the tasks up between production and maintenance (this division of labor is an active conversation point of the TPM team). Through autonomous maintenance groups, operators have greater involvement and say about equipment. As mentioned, TPM works when the operators begin to take responsibility for the equipment. As the sense of ownership spreads, autonomous maintenance becomes a reality.

The process must be implemented by all departments.

These departments include maintenance, engineering and tool/die design, and operations. Like many other programs of this type, TPM is a partnership of maintenance and production. The partnership will affect all the other stakeholders of maintenance. Their involvement is necessary for TPM to thrive. Every employee must be involved in TPM, from the workers on the floor to the president of the organization.

Pillars and Plinths of TPM

Based on the works of J. Venkatesh (see bibliography), and Imants BVBA (leading TPM consultants in Belgium), the traditional TPM has eight pillars. For our purposes, I argue that TPM has six pillars resting on four plinths (a flat block used as a base for something or a square block beneath a column, pedestal, or statue). Exhibit 3-1 summarizes this structure.

Why Plinths? The traditional TPM pillars are specific strategies to transform the plant to a lean and mean state. They are specific in that in different plants, different pillars will become more or less important. The pillars can be changed around, redefined for different outcomes. In one form or another, the plinths cannot.

Underlying the pillars are attitudes, systems, and procedures that are necessary for any manufacturing strategy. Whereas a manufacturer can run without Autonomous Maintenance, it cannot run without Safety, Health, and Environment security. The plinths are the basis of any business and the pillars are the basis of this particular approach to business.

PILLAR 1 — Jishu Hozen (Autonomous maintenance)

Autonomous maintenance is controlled by the TPM team, not outsiders. (The Japanese term Jishu Hozen also means independent in mind or judgment; self-directed.)

The TPM teams act based on established operating procedures, solve problems that develop, and seek to improve the overall effectiveness of the equipment. Effectiveness is measured by the metric OEE (to be discussed in a Chapter 6 in detail). Developing and training operators who can take care of small maintenance tasks will free up the full-time maintenance people to spend time on more technical repairs and PMs. The newly-skilled operators are responsible for upkeep of their equipment to prevent it from deteriorating.

Maintenance does not control an autonomous TPM team. This lack of control is one of the great strengths and weaknesses of the whole program. It is a strength because if the operators are close to the equipment and hands on, constantly monitoring the performance of the equipment, and keeping it clean, tight, and trim then other maintenance requirements will plummet. But if the TPM team is effective in name only, then the equipment will deteriorate at an increased pace, causing greater maintenance department intervention.

POLICY

- Uninterrupted operation of equipment or equipment available whenever needed
- Flexible operators to operate and maintain their equipment
- Eliminating the defects at source (in the equipment or even in the part made) through active employee participation, observation, and suitable   training
- Focus on Kaizen, lean, and small group activities

JISHU HOZEN TARGETS

- Reduce downtime (both scheduled and unscheduled)
- Reduce oil consumption of machines that use it
- Reduce process time (takt time or cycle time) by 50%
- Reduce variability in product output and quality
- Increase use of autonomous maintenance by 50% (inspection, cleaning, bolting, and lubrication / coolant / hydraulic)

STEPS IN JISHU HOZEN

- Preparation, training, and motivation of employees
- Review of O&M manuals to insure specifications are followed and update to field conditions and current OEM practices
- Initial cleanup, inspection, and lubrication of machines
- Take counter measures to reduce maintenance needed
- Recommend maintenance improvements that will make TPM activities faster, cheaper, easier, and more effective (projects that involve the maintenance department)
- Set standards for all activity
- Ongoing general inspection
- Autonomous inspection
- Standardization (everyone should do the same thing, the same way to the same equipment)
- Big question: How to manage autonomous maintenance and keep it going

PILLAR 2 — Synchronized Maintenance

Maintenance and production schedules are completely synchronized so that scheduled maintenance work occurs when the equipment is not in demand.

POLICY

- All production events that take the equipment out of service (model, size, color changes, replacement of blades, etc.) are shared with the maintenance schedule
- Maintenance requirements are published well ahead of time (52-week PM schedule)
- PM task lists are separated into non-interruptive activities to suit the window available

- Downtime due to scheduled maintenance reduced 75%
- Maintenance schedule reviewed and approved by operations
- Schedule miss due to operations down by 50%

PILLAR 3 — Proactive Maintenance

Trouble-free machines and equipment that produce defect-free products for total customer satisfaction is an excellent goal. This breaks maintenance down into three "families" or groups.

- Group 1: The Reactive group is Breakdown Maintenance; it is restricted to random events because any detectable deterioration would have been detected and restored before failure.
- Group 2: The Proactive group is Preventive Maintenance, Predictive and Condition-Based Maintenance, and Corrective Maintenance restoration of asset to like new condition resulting from inspections). We constantly evolve our efforts from reactive to proactive, using maintenance staff to help train the operators to better maintain their equipment.
- Group 3: The last group represents the Zone of No Maintenance Needed, based or Maintenance Prevention, Maintenance Improvement, Maintenance Elimination, and improved acquisition of better and more robust assets. Eventually we want maintenance-free designs.

POLICY
- Achieve and sustain availability of machines
- Optimum maintenance activity
- Reduced spares inventory needed to support assets

- Improve reliability and maintainability of machines

TARGET
- Zero equipment failure and break down
- Improve reliability and maintainability by 50%
- Reduce maintenance cost by 20%

PILLAR 4 — Product Quality

Forward-thinking organizations create customer satisfaction by delivering the highest quality goods or services. Focus on eliminating non-conformance in a systematic manner. We gain understanding of what parts of the equipment affect product quality; next, use that information to eliminate current quality concerns; and then move the focus to potential quality concerns. This process is analogous to the transition is from reactive to proactive maintenance (Quality Control to Quality Assurance).

POLICY
- Defect free conditions and control of equipments
- Focus on poka-yoke (foolproof or mistake-proof systems)
- Focus of prevention of defects at source
- In-line detection and segregation of defects
- Root Cause Analysis of any defects detected

TARGET
- Achieve and sustain customer complaints at zero
- Reduce in-process defects by 50%
- Reduce cost of achieving quality by 50%

PILLAR 5 — Continuous Improvement (CI) for Both Equipment and Processes

CI is a system of small improvements carried out on a continual basis involving all people in the organization. This system is the opposite of big, spectacular innovations. Continuous improvement projects generally require little or no investment. The underlying principle is that often a very large number of small improvements are more effective in an organizational environment than a few larger improvements. Such systems reduce losses in the workplace that affect our efficiencies. These activities are not limited to production areas; they can be implemented in administrative areas as well.

CI POLICY

- Practice concepts of zero losses in every sphere of activity
- Relentless pursuit to achieve cost reduction targets in all resources
- Relentless pursuit to improve overall plant and equipment effectiveness
- Extensive use of PM analysis as a tool for eliminating losses

CI TARGET

Achieve and sustain zero losses with respect to minor stops, measurement and adjustments, defects, and unavoidable downtimes. Aim to achieve 30% manufacturing cost reduction.

Tools used in CI:

- Lists of waste areas
- Lean projects
- Kaizan events
- PM analysis such as PMO
- RCA, 5 Why analysis
- Summary of losses

PILLAR 6 — Management of New Equipment

Proper design, commissioning, and testing will insure equipment can reliably make products to specification. Explicit and appropriate New Equipment Management procedures avoid the irrationality of saving a dollar on equipment to pay 10 dollars in maintenance and scrap. In the facilities field, good new equipment management involves choosing systems based on lowest life cycle costs, not lowest acquisition costs. Many of these items are based on the work of W.E. Deming.

POLICY

- Enough time to test-run equipment
- Enough time to design and build the equipment if it is custom-made
- Enough time and resources to properly install and then commission equipment
- Operator and operations involvement from early in the process
- Operator training and participation in start-up
- Maintenance training, particularly if there is new technology
- Latent defect analyses (run the machine over-speed to see what fails, then re-engineer it)
- Rebuild or re-engineer to your own higher standard
- Formal procedures for start-up and commissioning

TARGET

- Start-up problems reduced by 50%
- All master files in CMMS when commissioning is complete, including BOM (Bill of Materials)
- Initial PMs in CMMS when commissioning is complete
- Develop a standards book of the best components

- SOP (standard operating procedures) for operations and maintenance

The Plinths

The 4 plinths are the necessary foundation stones for successful businesses. The pillars are important for TPM and the plinths are important for continued success as an ongoing entity.

Plinth 1 — Safety, Health, and Environment

Without a safe place to work, the other pillars are just window dressing. For organizations to be sustainable over time employees, communities and the local environment have to be protected. Organizations must treat people and the environment with respect (at a minimum).

TARGET

- Zero accidents
- Zero discharges
- 100% recycling
- Zero health damage
- Zero fires, explosions

Plinth 2 — Training

All successful companies invest in training or risk losing their success. This plinth is important for long-term success. The staff consists of multi-skilled, revitalized employees, whose morale is high and who are eager to come to work. They perform all required functions effectively and independently. Operators are educated to upgrade their skills. It is not sufficient to know only "know-how"; they should also learn the "know-whys". The goal is to create a factory full of experts.

Policy:
- Focus on improvement of knowledge, skills, attitudes, and aptitudes
- Create a training environment for self-learning based on perceived needs
- Support OJT (on the job training) efforts with materials and aids
- Training curriculum / tools / assessments are conducive to employee revitalization
- Improve techniques to remove employee fatigue and make work enjoyable

Target:
- Achieve and sustain zero losses due to lack of knowledge / skills / techniques
- Aim for 100% participation in suggestion schemes

Plinth 3 — 5S

Without a physical location that is workable, production processes cannot thrive. Having the right stuff, in the right quantities, in the right place, and little else cluttering the place up helps people focus on the productive process.

Clean the place up! TPM starts with 5S, the much-vaunted Japanese process. 5S achieves a serene environment in the work place by involving the employees with a commitment to sincerely implement and practice systematic housekeeping. Problems cannot be clearly seen when the work place is unorganized. Making problems visible is the first step toward improvement.

SEIRI — SEPARATING

Seiri refers to the practice of going through all the tools, materials, etc., in the work area and keeping only essential items. Everything else is stored or discarded. This practice leads to fewer hazards and less clutter to interfere with productive work.

SEITON — ORGANIZE

The concept here is that "Each item has a place, and only one place." The aim is to arrange items so that each one is in easy reach, preferably near where it is used.

SEISO — SHINE THE WORKPLACE

This step involves cleaning the work place, for example no loosely hanging wires or oil leakage from machines.

SEIKETSU — STANDARDIZATION

Employees have to discuss together and agree on standards for keeping the work place, machines, and pathways neat and clean.

SHITSUKE — SELF DISCIPLINE

Considering 5S as a way of life can bring about self-discipline among the employees of the organization. The 5S mental attitude includes wearing badges, following work procedures, punctuality, dedication to the organization, etc.

Plinth 4 — Office TPM

The strongest body is just a shell without a mind controlling it. The mind is the office. They process orders, engineering, purchasing, and all the functions that are essential for the shop floor to work at all. The shop floor is only as good as the policies, systems, procedures, and overall effectiveness of the office.

Office TPM should be started after activating other pillars of TPM. Office TPM will provide improved support, productivity, and efficiency in the administrative functions. The goal is the same as shop floor TPM — identify and eliminate losses. This includes analyzing processes and procedures, usually increasing office automation. Office TPM addresses twelve major losses. They are:

- Processing loss
- Cost loss from areas such as procurement, accounts, marketing, and sales
- Communication loss
- Idle loss
- Set-up loss
- Accuracy loss
- Office equipment breakdowns
- Communication channel breakdowns, telephone and fax lines
- Time spent on retrieval of information
- Non-availability of accurate on-line stock status information
- Customer complaints due to logistics
- Expenses on emergency dispatches and purchases

Topics for Office TPM:
- Inventory reduction
- Lead time reduction of critical processes
- Motion and space losses
- Retrieval time reduction
- Equalizing the work load

- Improving the office efficiency by eliminating the time loss on retrieval of information, by achieving zero breakdown of office equipment like telephone and fax lines

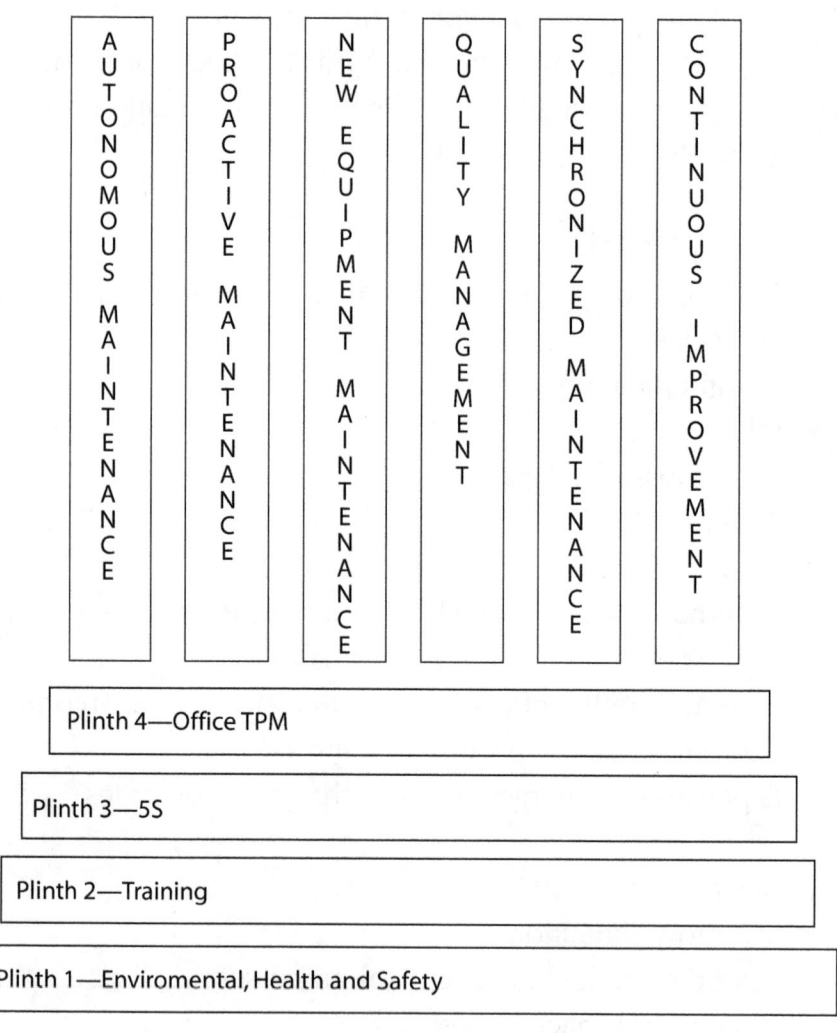

Exhibit 3-1 Pillars and Plinths of TPM

Chapter 4
TPM Basics

The Intention of TPM Is Engagement

All of the complexity, all of the Japanese words, and all of the pillars and plinths point to one underlying driver for TPM. That driver is engagement. The program requires and helps create an engaged work force. The point is that the Japanese have realized the power of an engaged worker. In general the West has not (although specific organizations have and are legendary). Successful programs are designed first to focus and then to encourage operators to start thinking about the whole process of production/operations.

Getting people to wake up and be engaged in their work is a challenge in all walks of life. There are certain jobs we are familiar with where being awake is an explicit goal. Think about the airplane mechanic or hospital inspector. We genuinely hope they are closely examining the item and not just glancing at it while thinking about the game last night or a fight with a roommate or spouse.

TPM transfers power (the ability to have a say about one's job) from management to production/operations. It gives significantly increased power to the operators. This new power changes people's minds about the areas in which they can have an impact. TPM works inside the culture of the organization to transform the relationship of people to the output.

TPM: The Basic Idea

TPM causes the operators to open their mind to all aspects of what the machine is doing. The specific techniques that help operators open their mind includes performing all of the routine preventive maintenance (e.g., cleaning, bolting, routine adjustments, lubrication, taking readings, start-up/shut down, and other periodic activities). In addition to these daily and weekly tasks, the operators are in center stage for problem solving concerning quality, output, and set-up reduction.

Operators are already busy and have little discretionary time and less free time. The key is what activities represent the best long-term value to the company for this valuable time. In fact, when Daniel T. Daley read an early draft of this book, he highlighted that the "objective is to use available operator time performing tasks that make the most cost-effective use of their time. Their (operations management) only difficulties are: How to populate available operator time with useful tasks and how to structure the available time in a manner that gets those tasks done in a timely manner."

One unfortunate fact is that when there is a great idea in business, the language is co-opted by management style faddists and applied as though it was a magic wand. Usually the breakthrough concept has antecedents that are necessary to achieve the great idea. Without the background work, the great idea is just a word.

The goal of opening the operators' minds is creating a feeling of equipment ownership for the operators. In management meetings, ownership is an almost meaningless platitude that actually means manipulation of the operators to feel ownership without a blip of support, resources, training, or even respect from management.

Once the operators become the "owners" of the equipment/area, then departments such as maintenance become valued advisors and specialists in major maintenance, major problems, problems that span several work areas, and as trainers. This also becomes true of other support functions such as metrology, process engineering, and quality.

This is a hands-on business! To TPM operators, their machines are like owning a 1967 Corvette. When you own one, you are hands on, cleaning, tightening, lubricating, loving inspection and true tender loving care. One of the tenets of TPM is that hands-on is essential for cleaning and close-in inspection. If the spirit of TPM is being followed, operators will be getting their hands dirty (and enjoying it). Generally the people who interact with the physical asset have an immense wealth of knowledge and information about how the asset is performing.

TPM Is a TEAM-Based Activity

Along with the emphasis on getting your hands on the asset is the focus on teamwork. Workers do not operate in a vacuum. This is a case where the whole is greater than the sum of the parts. Much more power can be derived from TPM teams than from its individuals. The team becomes the check and balance for TPM attitude. Operators report results and bring problems to the team. The team can see if the operators are acting consistently with the philosophy.

At a minimum, the TPM team (at one time or another) has members from the ranks of maintenance, machine operators, and set-up people. Others members from engineering, quality assurance, and even accounting or marketing can be added on an ad hoc basis. The members of the teams need to have adequate access to the training, resources, time, and downtime to accomplish their objectives.

Chapter 5
Maintenance

Maintenance

According to dictionary definitions, maintenance is "a series of activities designed to hold or keep something in any state." In concept, these activities do not have to be housed in a department. Anyone could conceivably be trained and assigned to these tasks. More than any other program discussed in the last 50 years, TPM assigns the tasks of maintenance to parties in different departments.

The Basic Activity of TPM Is PM

Let's examine a maintenance PM procedure and see how it applies to TPM. The following example is from the author's book *The Complete Guide to Preventive and Predictive Maintenance* (Industrial Press).

"Many of us start the day with a good cup of coffee from one of the increasingly common coffee shops. These shops feature a complex espresso machine. Almost all the tasks on the task list below are related to quality assurance. If you get a couple of bad cups of coffee you

Exhibit 5-1   Unit-Based PM Frequencies

Frequency	PM Routine for Espresso Machine Adapted from Programs of Espressoparts.com http://www.espressoparts.com/allabout/maint.html	Description
1. Q	Service filtration system (or outsource this item). Regenerate softener.	A water filtration system should be in place for most espresso machines.
2. W	Clean the group head.	Important — this is where the coffee comes in contact with the machine.
3. D	Back flush with water.	Back flush the machine for about 15 seconds. The blind filter will cause the water to pressurize, and help to clean out any accumulations of coffee grounds and oils that may have formed.
4. W	Back flush with Purocaf or other NSF approved detergent. Important note: Do not back flush piston machines! Instead, just replace the screens and gaskets on a regular basis.	Afterwards it is important to remove the porta-filter and run the group again to rinse out all remaining detergent and back flush several more times with water only. In addition to rinsing, one or two shots of espresso should be extracted through each group to "re-season" the machine.
5. W	Soak porta-filters and screens in detergent (after back flushing).	Follow dilution instructions with very hot water. Be sure however to rinse porta-filters well before re-using.
6. D	Clean group gaskets.	Cleaning is best accomplished using a specially designed group cleaning brush and hot water to vigorously scrub around the sealing surface.
7. D	Purge and clean the steam wands.	Use warm soapy water and a non-abrasive cloth to remove all milk residues.
8. D	Examine the steam wands for cracks or signs of the chrome plating flaking off.	Either condition would require immediate replacement of the wand.
9. D	Remove the drain tray and clean the drain cup.	Pour a pitcher of hot water into the drain cup to help rinse accumulated coffee grounds down through the drain hose.
10. A or on-condition	Replace group head shower screen.	Even with regular back flushing, the group head shower screens must be replaced periodically.

will cross that shop off your list. Note that even in a small-scale retail operation, an economic analysis of this task list is still possible."

If we owned hundreds of coffee shops it would be important to look closely at each task and see what is the optimum frequency and who is the logical person to be responsible for that task consistent with high quality at the lowest cost. We might want to abandon the calendar-based system in favor of a utilization (shots of espresso) based system. Other factors might include the types of coffee that are popular, weather, water composition, and skill of the operators. Cutting 10 minutes a day from the PM routine (without any quality sacrifice) in a hundred stores might be worth over 4000 hours a year (that could be available for customer interaction)!

The unit-based PM frequencies for this machine, summarized in Exhibit 5-1, are (A-Annual, Q-Quarterly, M-Monthly, W-Weekly and D-Daily):

Let's accept that this PM list is economical and soundly engineered. The question is who should do each of these tasks and when should they be done?

It is easy to see that the annual and quarterly tasks need specialized skills. We should either have a high-skill, roving PM person (who may also handle breakdowns and service problems) or a contractor handle them. The daily tasks just as clearly need to be done with someone right there on-site to minimize travel time. The weekly tasks require chemicals (detergent) so we might have a senior operator do these.

In the daily cases, we would have a complete class for the operator in what to do, how to do it, and why it is done. This would be a great application of video training along with narrated Power Point, ending with a test. The culmination of the training would be performance of the tasks under the eye of a senior operator.

If we took the next step toward TPM in coffee shops, we would organize the operators to look at the whole productive function and try to optimize the process to increase throughput and quality. They would use the 6 or 11 production losses as pointers (thoughts and discussions about these losses — among the core concepts of TPM — are found in detail in Chapter 6). Of course, active measurement of the current state of the losses would be essential.

The D-Daily and W-Weekly items might be printed on a placard and mounted directly on the machine. The placard must be mounted where it can be seen.

Maintenance tasks have always been seen as activities determined by an engineering review of the needs of the operating asset. TPM adds the human requirements of operators running the machines day in and day out to the engineering requirements. For the best work, the operators have to be involved in the care of the machine, in designing improvements, and in being consulted when any aspect of the machine is to be impacted.

At a minimum, TPM works with other good maintenance practices to achieve the overriding goal maximal output. By tracking and eliminating wasted machine time, stoppages, and quality problems, we see that asset utilization goes up, quality output increases, and profit improves.

Industrial assets have a widely assumed amount of acceptable uptime that regulates the overall amount of output. This unspoken ceiling represents a lost opportunity amounting to millions of dollars of profit. TPM techniques challenge this underlying assumption in two areas: ideal production rate of quality product and ideal levels of downtime.

TPM establishes what the true production output ceiling is for each set of assets. TPM attacks all the areas that can impact both actual uptime and production rate to achieve that ceiling.

The potential is that, after the TPM teams start to have an impact, the false ceiling gradually rises to almost match the true ceiling. The true ceiling might rise too as the bottlenecks are worked out.

Although the basic driver for TPM is waking up the involvement and creativity of the operators, one of the basic activities of TPM is PM (Preventive Maintenance). PM tasks are among the core activities of the TPM teams.

PM Is...

To clarify, PM is a series of tasks that:

Prevent —
Postpone —   Failures
Predict —

Another way to say this is:

1. PM extends the life of an asset. For example, greasing a bearing of a motor or gearbox will extend its life.

Or-

2. PM detects that an asset has had critical wear and is going to fail or break down. For example, an inspection shows the drain hose on the left of Exhibit 5-2 has disconnected itself from the drain. This

discovery allows you to reattach it while it is a minor problem, well before a catastrophic breakdown develops (e.g., dripping condensate causes rust through the bottom on the unit or contaminates expensive products). If this is a common reoccurrence, the TPM team should work with maintenance and/or engineering to get to the root cause of why the hose came loose and fix it permanently. (See more on Root Cause Analysis in Chapter 7.)

Inspection

Common tasks associated with PM and TPM teams — what TPM operators actually do — include inspection tasks. These tasks are the low tech, highly skilled application of human senses including seeing, feeling, smelling, hearing, and tasting (only occasionally such as wine production) for problems. In Exhibit 5-2, a visual inspection is all that was needed. As part of the TPM team, operators are usually in the best position because they see the machine in every state and are totally familiar with its ordinary sounds, behavior, etc. For example, look for leaks in the hydraulic system; listen for bearing chatter. The TPM-trained team member could repair such minor faults on the spot. Paul R. Casto observed that "Another advantage is that the inspections also activate the operator's knowledge base. The operator is an expert system that has sensors (smell, touch, hearing, seeing, etc)."

Predictive Maintenance

Predictive maintenance is inspection with some help from technology. It can be vibration analysis, infrared scanning, or even mega ohm readings on a motor winding. For example, scan all electrical

Exhibit 5-2 A Disconnected Drain Hose

connections with an infrared detector. Normally the TPM team is directly involved in taking readings, but is less involved in the analysis of sophisticated predictive maintenance scans or reports.

TLC

The next three activities comprise **TLC**.

Tightening

Looseness is the second biggest cause of machine breakdown. Bolting includes looseness, missing fasteners, misapplied fasteners, and wrong fasteners. For example, tighten anchor bolts. The TPM team is totally involved in bolting issues.

Lubrication

Basic PM includes lubrication. This is a critical area, and yet one where training is often slip-shod or non-existent. Investigate automatic oilers, which present significant opportunities in this area. For example, add 2 drops of oil to bag stitcher. The TPM team is totally involved in lubrication issues.

Cleaning

A study in the Japanese Society for Plant Engineering showed that 53% of all breakdowns in factories were caused by dirt and bolting problems. For example, remove debris from machine. The TPM team is totally involved and responsible for all cleaning. Teams are trained in detailed cleaning, analyzing where contamination is coming from and fixing it.

Additional Tasks

Operate

For all equipment, the TPM team is concerned with correct operation to eliminate all sources of human errors. In addition to error proofing (poka-yoke), TPM can develop standard practices to operate infrequently used equipment. On some equipment that is used infrequently, observing the operation is the only way to insure it will work when needed. Examples include running an emergency generator for an hour each week and advancing the heat control on injection molding machine until the heater activates. Of course, the operators and the TPM team are both responsible for operating and observing the equipment.

Improvement of conditions is called maintenance improvement (also called maintenance prevention)

The TPM team should be spending a good deal of time improving conditions. The goal is to spend as little time on equipment service as possible, while getting the job done. That includes sourcing contamination and eliminating it, making it easier and quicker for the TPM operator to lubricate, eliminating some routine maintenance activities, and making tasks easier.

Adjustment

The goal of adjustment is producing quality output every time. No scrap should result from out-of-adjustment situations. In addition to impacting quality, maintaining the life of the asset is also important. Many components fail because they are allowed to come out of adjustment — such as belts, and limit switches. For example, adjust the tension on the drive belt. Ultimately a TPM team goal is to eliminate or standardize as many adjustments as possible. The TPM team is usually trained and responsible for adjustments.

Set up and changeover of machines

Part of the tasking may be the set up and changeover of machines. This step might include changing tooling or product color, size, and specification. (There might be a set-up team, but the concept is to simplify and standardize set-up and bring this into TPM.)

Take readings and record measurements

Measurements and readings can detect deterioration before it is obvious by other means. Many maintenance events follow unusual readings. Other events follow a slow decay in a key parameter that could have been detected by readings such as boiler failure, air compressor failure, and filter changes. For example, record readings of amperage. The TPM team is accountable to take readings.

Communicate

One activity that is often overlooked is the responsibility to communicate. Timely, accurate, and complete communication of corrective repairs needed and other issues (that are beyond the expertise of TPM operators) with maintenance and other groups is essential for a successful program. The CMMS and the Work Order system are the keys to the maintenance department communications. In Chapter 9, we will visit this issue in depth.

Scheduled replacement of minor components

Minor components include indicator lights, some belts, and other consumables. This step is also called PCR (planned component replacement). For example, remove and replace a urethane bolster every month. This area is not usually the responsibility of TPM unless the item is small and can be replaced in 30 minutes or less without significant trade knowledge. TPM teams will certainly replace cutting blades, etc. PCR is used effectively by airlines to produce ultra-high reliability.

Interview Other Operators

This step is especially important if one operator is TPM and others are not. Ask questions and build a relationship of mutual respect. The operator is the closest to the action. For example, ask the operator how the machine is running, if there are any unusual noises, or if excessive heat is being generated.

Analysis

Review the PM and repair history. Look for areas of possible improvement. The TPM team is involved in analysis, but is not always accountable for the outcome. Certainly issues to analyze would come from the TPM team.

Lead focused projects

Although focus projected are often facilitated by engineering or maintenance professionals, TPM members would be logical candidates for that effort. One such project is the 1-point lesson.

One Way PMs are Built: Team Exercise

After you've looked at the O&M manual for clues and reviewed all the existing PM documentation already developed, the next place to review is your CMMS. If your CMMS database is not complete or is invalid, the memory of your long-time employees can be substituted. The key is to look at the current failure modes. A good place to start is to list the most common, most expensive, and most dangerous failure modes.

Make a list of the failures and make a list of the PM tasks that would eliminate each of these failures. Evaluate each task for its suitability for TPM.

Chapter 6
OEE
(Overall Equipment Effectiveness)

Measuring Equipment Effectiveness Is an Essential Part of TPM

Thought exercise:

List all of the potential reasons that a line cannot produce the maximum amount of product, given its rated production and the number of minutes it runs.

Data is the key

Many organizations do not or cannot capture accurate information about run time, slowdowns, minor stoppages, and defects. TPM relies on good record keeping in the areas of loss.

OEE for a Donut Shop

Let's say we make donuts. We have an asset (donut machine assembly line) that routinely makes say 20,000 dozen donuts a week.

Big pile of donuts
(The ultimate goal!)

Improvement
(Available from TPM)

Small pile of Donuts
(What we already have)

Ideal Production or
100% OEE

After TPM 67% OEE

Current Production
50% OEE

Exhibit 6-1 Comparing Current and Target Production Levels

The ideal production rate is say 40,000 dozen a week. That amount requires everything to be perfect. With TPM, maybe we can increase our production to 25,000 dozen by making the equipment run at its name plate speed and by reducing quality problems (Exhibit 6-1). The ratio of our current production level to our calculated ideal production level is called OEE (20,000/40,000 = 50%).

Different OEE Models

TPM can be summarized as attention to and elimination of any losses to production by the operators and their team. Different TPM experts have different lists of the losses that follow.

6 Losses of Nakajima

Below are the 6 losses first written by Nakajima in his seminal works Introduction to *TPM* and *TPM Development Program*. They fall into three categories: downtime, speed losses, and defects.

DOWNTIME

Equipment failure requires maintenance assistance, but may be prevented with the use of appropriate preventive maintenance actions, developed and applied operating procedures, and design improvements. In this model, scheduled downtime for PM activities or shutdowns are subtracted from loading time. Most important, equipment failure requires an improvement effort that should result from a successful partnership between production and maintenance.

1.   *EQUIPMENT FAILURE FROM BREAKDOWNS*

Equipment failure causes production downtime. This is the biggest element that is directly the responsibility of maintenance. With TPM, the first line maintenance activity to avoid breakdown is transferred to operations. Proper design and use in the first place insures reductions in breakdown-related downtime.

2.   *SETUP AND ADJUSTMENT*

This refers to loss of productive time between product types, and includes the warm-up after the actual changeover. Changeover time should be included in this loss opportunity, and it should not be part of the planned downtime. The stated goal is called often known as SMED (Single Minute Exchange of Die). This allows up to 9 minutes for set-up. Adjustments are simplified or eliminated from the system. Overall re-engineering to reduce these exposures is expected.

SPEED LOSSES

3.   IDLING AND MINOR STOPPAGES

Small stops are typically less than 5–10 minutes; they are usually for minor adjustments or simple tasks such as spot cleaning. They should not be caused by logistics snafus. They could be due to abnormal operation of sensors, blockage of work on chutes, etc. These slowdowns are tracked and analyzed to see what is really happening. Analysis of root causes and of process is ongoing until the system no longer has losses in these areas.

4.   REDUCED SPEED DUE TO DISCREPANCIES BETWEEN DESIGN AND ACTUAL SPEEDS

Design speeds are reviewed and actual speeds are observed. The comparison, if unfavorable, initiates a design and engineering review. Examples include machine wear, substandard materials, operator inefficiency, and equipment design not appropriate to the application.

DEFECTS

5.   PROCESS DEFECTS DUE TO SCRAPS AND QUALITY DEFECTS TO BE REPAIRED

Losses during production include all losses caused by less-than-acceptable quality after the warm-up period. Quality problems are not tolerated. Deep analysis is undertaken until these losses approach zero.

6.   REDUCED YIELD FROM START-UP TO STABLE PRODUCTION

Losses during warm-up include all losses caused by less-than-acceptable quality during the warm-up period. The production process is tracked and watched for start-up problems. Stable production should follow start-up very closely.

11 Losses of Bob Williamson

Exhibit 6-2 shows additional variables in calculating effectiveness. Please note how the calculations for the overall equipment effectiveness are related to the 6-loss scenario. The original chart of losses according to Bob Williamson from TPS included 11 losses.

Exhibit 6-2
11 Losses of Bob Williamson
Availability losses
Planned shutdown losses
No production scheduled (1)
Planned maintenance (2)

Downtime losses
Breakdowns and failures (3)
Changeover (product, size) (4)
Tooling or part changes (5)
Startup or adjustment (6)

Performance efficiency losses
Minor stops (jams, circuit breaker trips, etc.) (7)
Reduced speed, cycle time, or capacity (8)

Quality losses
Defects/rework (9)
Scrap (10)
Yield/transition (from changeover, startup/adjustment) (11)

16 Losses

With the passage of time, others added more losses to the above list. Each organization has its own classification of losses. Exhibit 6-3 shows one such classification listing 16 types of losses.

Exhibit 6-3   16 Losses

Breakdown Set-up and adjustment Cutting blade loss Start up loss Minor stoppage / idling loss Speed loss Defect rework losses Scheduled downtime loss	Losses that impede equipment efficiency
Management loss Operating motion loss Line organization loss Logistic loss Measurement and adjustment loss	Losses that impede human work efficiency
Energy loss Die, tool, jig, fixture breakage loss Yield loss	Losses that impede the effective use of resources

Immants Classification of Losses

Immants BVBA also includes losses in materials and labor. This is consistence with their focus on Lean Maintenance (discussed in depth in a Chapter 7). Traditional TPM was part of the bigger production or operations system (TPS, as mentioned). The materials lost would be addressed in the Lean practices while the production was addressed by the TPM practices. Still, this categorization can be used to help form a complete picture (Exhibit 6-4).

Exhibit 6-4 Immants Classification of Losses

Labor Losses	Cleaning and checking
	Waiting for materials
	Waiting for instructions
	Waiting for quality hold

Material Losses	Material yield
	Energy losses
	Consumable losses

Let the Confusion Begin

As you can see from the examples, there are several versions of OEE. Practitioners of TPM follow one method or another. Some feel that the system they use is the best and are quite outspoken on their choice. In papers and articles, the different OEE calculations may not be cited explicitly, leading to even more confusion.

If there is a truth in this, it is that different industries need different OEE metrics. Even different business situations require different ways of looking at the data. For example, as this is written, many plants are running partial schedules. The SMRP (Society of Maintenance and Reliability Professionals) has weighed in on this issue. Their definition removes unscheduled down time due to no demand (SMRP: June 30, 2009 Metric Publication) from the loading hours.

Bob Williamson, one of the original TPM enthusiasts, says that OEE has created some confusion and even bad blood. OEE percentages became a metric to compare current equipment performance to world-class performance. For example, the measure of 85% equipment effectiveness became known as "world-class OEE." This "one size fits all" is a trap.

Once used as a benchmarking score for world-class, OEE became used as a way to compare one piece of equipment to another, even though the equipment performed different functions in a different process, or even in a different plant. Once this basic calculation became more widespread, OEE started being used to specify Overall Plant Effectiveness (OPE) by using an aggregate score for all equipment in the plant.

OEE and now OPE are widely used to compare current levels of maintenance effectiveness and equipment performance to world-class levels. The problem is that these measures are used as a club to punish those whose OEE (or OPE) slips. Such uses provide inaccurate, unfair comparisons; they are a gross misuse of the original purposes of OEE. OEE is ideally a tool to uncover barriers to ideal production levels, not as a crude whip to make people work harder.

TPM standards

Availability	>90%
Performance Efficiency	>95%
Rate of Quality Parts	>99%

Examples of OEE

Example 1

Assumptions:

Running 70 percent of the time (in a 24-hour day)

Operating at 72 percent of design capacity (flow, cycles, units per hour)

Producing quality output 99 percent of the time

When the three factors are considered together (70% availability x 72% efficiency x 99% quality), the result is an overall equipment effectiveness rating of 49.8 percent. Here is how that breaks down visually:

Availability 70%	30% Downtime loss
Performance 72% of 70% = 50.4%	28% Speed Loss
Quality yield 99% of 50.4% = 49.8%	1% Quality Loss

Example 2

Given on the following page are the details that contain shift data to be used for a complete OEE calculation. They are compiled into the three traditional contributing factors of availability, performance, and quality.

Case Study: Measuring Equipment Effectiveness

Greenfield Manufacturing is one of the leading producers of pipe hangers. They service the electrical and plumbing trades with hangers from 1/2" EMT to 8" cast iron pipe. They are a single shift, 5-day operation. All machines are shut down for periods of 30-minute lunch, 15-minute morning start-up, and 15-minute evening shutdown.

Shift Data	Calculation Data	Calculation
Shift length	480 minutes	Loading time 480 – 60 =
Short breaks and Meals	2 @ 15 minutes	420 minutes
(machine off)	1 @ 30 minutes = <60 minutes>	
Downtime (due to change<	50 minutes>	Availability 420 – 50 =
over and set-up)		370 or 88.1%
Design rate	70 pieces/minute	29,400 pieces
Pieces actually produced	20,225 / 29,400	68.7%
Rejected pieces	503	19,722 / 20,225 = 97.5%
OEE	59%	.881 * .687 * .975 = .59

Planned production time = Shift length less breaks = 480 – (15 + 15 + 30) = 420 minutes

Operating time = Planned production time – Downtime = 420 – 50 = 370 minutes

Availability = Operating time / Planned prod time = 370 Minutes / 420 Minutes = **0.881** (88.1%)

Design output = Pieces produced / (Design rate x Operating time) = 420 * 70 = 29,400

Performance = Actual output / Design output = **.687** (68.7%)

Good pieces = Pieces produced – Reject pieces = 20,225 – 503 = 19,722 pieces

Quality = Good pieces / Pieces = 19,722 / 20,225 = **0.975** (97.5%)

OEE = Availability x Performance x Quality = 0.881 x 0.687 x 0.975 = .591 (59.1%)

They operate punch presses and small press brakes ranging from 12-to-100 tons in semi-automated to fully automated modes. They have their own tool shop and rebuild presses to their own spec. Automation is a high priority. They use a strategy of minimizing setups by specializing presses so that changes in setup occur only once a week (when they need the press for an unusual size pipe hanger).

4" RISER CLAMP

This study concerns their new 4" riser clamp tooling and setup. Four-inch riser clamps are made from 3/16 X 1 1/2" milled steel. It is shaped with a hump in the middle between two ears so that the body of the clamp will hold 4" cast iron pipe securely and the ears will hold up the pipe at each floor level. The clamp is made on a modified press brake in progressive tooling with three stations (form, punch, and cut-off). The parts are conveyed and then dropped into boxes and taped closed. It is fully automated.

The press break (machine number is PB1), located in the Heavy Metal Department, is one of the slowest presses in the entire factory, rated at 20 strokes per minute. When we timed the press we found it was running at 18 strokes a minute (one piece was made per stroke). The maximum output at 18 strokes per minute is 37,800, with a lost output of 4200.

In a typical week, there are only 4 hours of down time from material changes (end of coils) and 3 hours a week lost to minor jam ups and the occasional setup. Because this is a rough-type part, quality problems are few and far between and are usually related to start-up, problems in the incoming steel, or coil ending losses. They estimate that the average reject rate from all sources is 25 pieces per day. The average production for the previous 30 working days was 6275 pipe hangers/day.

CALCULATIONS

Ideal production is (this is progressive tooling with forming, hole punching and finally parting with 1 piece coming off the press per stroke after the 3rd stroke on each new coil)

8 hours/day * 5 days/week * 60 minutes/hour * 20 strokes/minute = 48,000 strokes/pieces/week

6275(actual pieces per day from production reports) *
     5 days = 31,375 actual pieces/week
The gross OEE is then = 31,375/48,000 = 65%

There is a discussion in the field if the breaks and lunch should be included in the loaded hours (as they are above in the ideal number). If we remove the breaks and lunch from the loading time, the ideal drops to (same asset as before where 1 stroke equals 1 piece after the coil is started):

7 hours/day * 5 days/week * 60 minutes/ hour * 20 strokes/minute
     = 42,000 pieces/week
The gross OEE improves to 31,375 / 42,000 = 75%.

What we want to do is to determine exactly where the losses are coming from. Losses can only be impacted on the level that they occur. We must know where the problems are to know where the team has to look.

---

* I say cheated because, as you will see later in this section, real OEE requires knowing the exact source of the losses from the 48,000 ideal

Let's summarize what we know:

- The machine is running below speed at 18 strokes a minute. As a result, it is producing about 4200 fewer pieces a week.
- In a typical week, we lose 4 hours from coil ends (change material coils). Another way to look at this is we lost 4800 pieces during the time we were resupplying the machine.
- We lose about 3 hours a week on average due to jam ups and minor stoppages. Another 3600 pieces lost.
- Finally the company estimates that average reject rates from all sources are 25 pieces per day or 125 pieces a week.

Speed losses 42,000 − 37,800 = 4200 42,000/4200 = 10% loss	10% speed loss

Material issues 4 hours/ 35 hours = 90.1% 90% * 90.1% = 81%	9.9% loss

Jam ups, etc. 3 hours/35 hours = 91.4	8.6% Loss

Quality yield 42,000/125 =.03 = 99 .7% 74% * 99.7% = 73.8%	.3% Quality Loss

From charts of this type (adapted to your actual losses), the TPM team might first look at why the press is running below name plate speed. Fixing that problem will give us the biggest boost in OEE. The team will be looking at **all** the losses and over time fix what it can.

Chapter 7
TPM Activities

Return It to Like-New Condition

The goal of TPM activity is to maintain an asset and, when TPM is installed, return it to like-new condition (assuming the asset performed well when new). If there has been deterioration, then the operator will choose the appropriate action, such as replacing a worn part or writing up a corrective maintenance work order.

Much time is spent inspecting the assets by looking closely, touching, and listening to the operation. OEE and control charts tell the operators numerically that everything is as it should be. Once something goes amiss, the operators (and their team if indicated) jump into action. The action could be to conduct a 5-Why exercise; if the problem is more complicated, they might do a full RCA (Root Cause Analysis).

They will classify the loss, look closely, and try to understand the condition that changed. Once they come to a conclusion, they initiate appropriate action.

Mini-Manual for Operators

Operations need manuals of direct information for operators and competencies (Exhibit 7-1). It could be organized around "everything you ever wanted to know about your punch press." Everyone new to the operator group would be indoctrinated into the manual and use the manual as their bible in matters of the machine. This manual encompasses the standards for operations and maintenance of the asset. It must be kept up-to-date with new information and approaches. Any changes, updates, improvements or clarifications to the manual are vetted by the TPM team. After vetting, the changes are transferred to the master. At the same time, we encourage the operators to actively use it daily.

The operator's mini-manual (this might be a subset of the complete manual) would support and reinforce the lessons learned in the TPM training. Some of the topics covered would be:

- Glossary of all the terms and words for maintenance and operations of the asset
- Index, tabbed and logical organization making it easy to find things
- Description of the production program and the goals of all those activities
- Short descriptions of major concepts
- Background on what makes up OEE (sources of the losses that are the responsibility of the operator); how the metric is actually calculated (no secrets)
- How the machine works inside with pictures
- Standard operating procedures of the machine (one best way)
- Abnormal operation and what to do

- Trouble shooting for operators
- Defects observed by teams in the past
- What are the TPM tasks designed to do
- TPM task lists by frequency
- Detailed readings, torque settings, grease types, cleaning materials, and techniques

One area that could be expanded upon is the SOP (Standard Operating Procedures).

These procedures should be easy to understand and be organized to be easy to use. Because the person is trained with them, they should be good training aids. Readings and measurements should be based on actual standards.

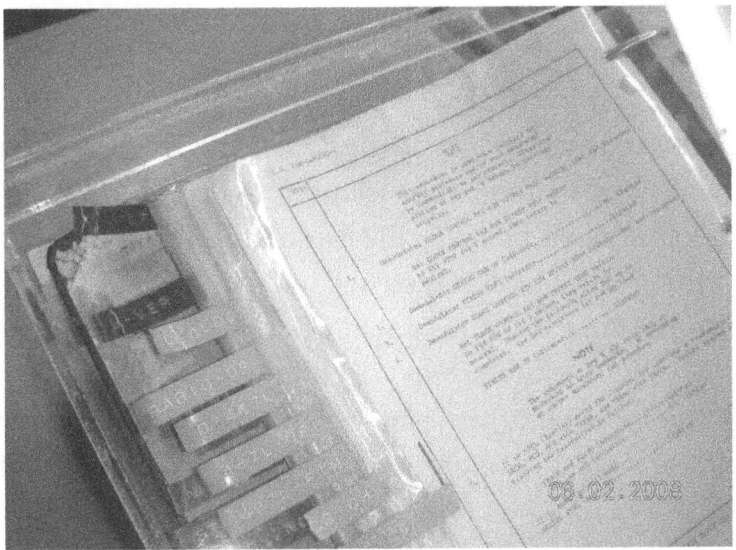

Exhibit 7-1 SOP for a Titan Missile silo from Titan Museum

The SOP would include all materials used, frequency, and cost data. Where possible, use pictographs (iconic pictures), charts, and other visual reminders.

Tasks (Getting Down to the Nitty-Gritty)

Tasks are reminders of what to do, how to do them, and with what materials and tools. They are not a substitute for alertness and sensitivity of the operator, but should be taught as a basic practice (for example, in sports, shooting baskets and running laps are practice for the game). As a practice, athletes want their bodies to move into the groove and be completely conscience without thought. That is the goal of TPM tasks; to look at everything and see if anything is amiss, all without thought.

Explicit versus Implicit Tasks

Implicit tasks require a good deal of experience or knowledge beyond what is written about the task itself. A task like "Go PM the air compressor" is implicit because the operators would be expected to know everything to do when they set out. An explicit task does not require the operators to know anything (or very much) outside the given task. So if we tell them to "top up the oil," we would add the type of oil, the location of the topping up line, and even instructions to clean off the oil cap and area before the task is performed with a new shop rag.

Implicit: Make sure the blade is set properly for clean cutting.

Explicit: Insert feeler gauge A into blade opening; it should fit snugly. Insert gauge B into blade opening; it should not fit. Include a picture of the blade opening, all appropriate safety information, and instructions of what to do if the gap is too small or too large.

TPM tasks are always explicit. It is essential that the tasks be specific and thought through. This is the ultimate planned job where all material is called out and the job steps are fully detailed, with any external reference included. Implicit tasks require background knowledge. Explicit tasks can be done with the information on the task list (although training is done too).

Good TPM Tasks Have Some Things in Common

Should be done daily or weekly: Because the operators have not committed themselves to a lifelong interest in maintenance, we want tasks that are frequent so they don't forget them.

Task can be well defined: This goes along with the tasks being explicit. The best tasks are well defined and could (in theory) be done by anyone.

Can be drawn, photographed or sketched with pictographic instructions: Pictures solve the dual problems of illiteracy and differences in language. Pictures are great as long as they are not too cluttered with extraneous detail (Exhibit 7-2a)

Exhibit 7-2a Can be drawn, photographed or sketched with pictographic instructions

Often iconographic drawings (simplified) are even better (but more expensive). Even people whose language is different than yours can understand the pictograph (Exhibit 7-2b).

Exhibit 7-2b iconographic drawings (simplified universally understood drawings)

Will give an excellent ROI in terms of uptime: Because they take time away from the production day, all TPM tasks need to be economically justified in terms of reduced repair time and cost, improved quality, or increased output.

Require basic tools: Tool use should be managed and minimized. In some cases, the tool can be safely chained to the machine. In others, a small toolbox should be provided. Proper tool use is a part of the TPM training.

Local materials: Parts and consumables should be stored near the machine and should not require excessive travel or special handling equipment beyond what the operator uses daily.

Do not expose the operator to excessive hazard: In some plants, operators are exposed to hazards and have been given specific training and PPE (Personal Protective Equipment) to mitigate their risk. Follow the same model for TPM tasks.

As the teams get trained and have some experience under their belts, they can take on rectifying faults, doing short repairs, and detecting impending failures.

Possible tasks for TPM-trained operators:
- Close in inspection of the asset using touch, sound, visual inspection, and smell
- Write a work order, notification, or work request
- Take meter and gauge readings
- Write readings and other information into log
- Take vibration readings
- Listen to device with ultrasound equipment
- Replace any items that are damaged (if small, safe, local, and within training)

- Detect and replace worn items (if small, safe, local, and within training)
- Fill in work order for repair and replace activities
- Predict failures from inspection
- Hands-on cleaning asset
- Cleaning area around the asset including above and below
- Cleaning lubrication points and areas
- Lubrication
- Hands-on bolt tightening
- Maintenance improvement (reduce maintenance needs by elimination of the source of the problem)
- Bring into adjustment improperly set items
- Participate on RCA, RCM, Kaizan, 5S teams
- Identify improvement opportunities
- Lead and participate in focused improvement projects
- Implement change for the better
- Gather and catalog the equipment's history

TLC (Tighten, Lubricate, Clean)

In maintenance, TLC — tender loving care —means tighten, lubricate, clean. TLC is a subset of PM and the key to TPM. Keeping equipment trim and clean will extend the life and reduce the level of unscheduled interruptions. This approach is appropriate for all maintenance departments, even those with no support from top management or maintenance customers. **TLC is the single simplest way to reduce breakdowns.**

Yet the climate seems to be against TLC. As firms experience downsizing and de-staffing, one of the first services to go is TLC. When we read the latest trade journals and listen to the latest papers at conferences, we hear and read that time-based (or interval-based) PM is obsolete. At a recent international maintenance conference, there were 35 papers or sessions presented; not one of them spoke about improved TLC.

Yet studies find again and again that dirt, looseness, and lack of proper lubrication cause a bulk of the equipment failures. TLC is the core of TPM's increased reliability. The examples below are from *TPM Development Program* by Nakajima.

- One company found that 60% of its breakdowns were traceable back to faulty bolting (missing fasteners, loose or mis applied bolts).
- Another company examined all of its bolts and nuts; it found 1,091 out of 2,273 (48%) were loose, missing, or otherwise defective.
- The JIPE (Japanese Institute of Plant Engineering) commissioned a study that showed 53% of failures in equipment could be traced back to dirt, contamination, or bolting problems.
- Effective TLC can impact other costs. One firm reduced electric usage by 5% through effective lubrication control.

TIGHTEN (ALSO CALLED BOLTING)

Loose or missing bolts are a major source of breakdowns. Even a single missing or loose bolt might cause a failure. In most cases, the looseness contributes to vibration, which increases looseness even more. In electrical joints connected by the pressure of a bolt, looseness is usually the result of thermal expansion and contraction. The space created by looseness harbors oxidation, which increases resistance, which expands and contracts the joint, which causes more looseness. In other words, loose bolts beget loose bolts. In a review of an early draft of this work, Paul R. Casto offered "The worst way to tighten is to do it with so much torque that material damage occurs and then the fastener will never stay tight."

The misapplication of one nut cost an air charter company $9,600,000. In 1990 a plane crashed in the Grand Canyon. The nut holding the propeller on a small tour plane came loose, causing the propeller to fly off. The jury awarded $9,600,000 for negligence in

and cocaine, partly from an abusive home.

Mr. Zeitvogel, who was sentenced to life in prison for the 1981 stabbing, was originally jailed at the age of 18 for his role in a 1974 rape and bur-

smile ... only in his fina. "I'm sorry to the family for t. I have caused," he said. "I hope t. closes the chapter on this."

12/12/96   NY TIMES

Loose Screw Caused Stuck Hatch on Shuttle

CAPE CANAVERAL, Fla., Dec. 11 (AP) — A loose screw caused a hatch on the space shuttle Columbia to jam and kept the crew from taking two space walks, NASA said today.

The screw fell from its hole and became embedded in the hatch's gears, probably during liftoff, said Bruce Buckingham, a spokesman for the National Aeronautics and Space Administration. A second screw in the braking mechanism for the hatch handle also came loose but remained in place, Mr. Buckingham said.

Engineers discovered the loose screws after removing the gear box from the hatch and taking it apart late Tuesday. It is not yet known whether technicians had properly installed the quarter-inch screws, Mr.

Buckingham said.

The shuttle returned to the Kennedy Space Center here on Saturday after an 18-day mission. Two space walks intended as dress rehearsals for space station construction had to be canceled because of the stuck hatch.

NASA is debating whether to remove the gear boxes on the hatches inside the shuttle Atlantis on the launching pad, to make sure all the screws are tight.

Such a job would probably require overtime work over the Christmas holidays in order to get the work finished in time for the Jan. 12 launching, Mr. Buckingham said.

DO NOT FORGET THE NEEDIEST!

Exhibit 7-3 A very expensive loosened bolt

deductible insurance policy. If a main nut holding a propeller can be overlooked, what is the chance that you have nuts working their way loose, even as you read this section! It's not only planes but sophisticated space gear. Exhibit 7-3 shows another example, this one involving NASA and the space shuttle.

The people cleaning the machines have the best chance of detecting this failure. As they touch and look at the machine, loose bolts should shout to them. The easiest technique is to scribe a line on the nut and the machine frame when the nut is tightened correctly. This scribed line will stay lined up (a single line) as long as the nut doesn't move.

When equipment is first engineered, the rules of good bolting should have been followed. Much of the process of maintenance is correcting mistakes or deviations from good engineering practice. Many rules concern the size, pattern, torque, and type of fastener. Other rules include head location (nut is accessible), use of lock washers, use of flat washers, and bolt length. In most facilities there are no well-known standards for tightness, though that would be helpful for task lists with tasks such as "check base bolts and tighten if loose."

Bringing equipment to specification is sometimes a lengthy job. New vehicles in a truck fleet are brought in after 1,000 miles to tighten everything up. This short run-in period gives the bolts a chance to seat. This same strategy would be useful after factory rebuilds or when doing work in buildings. Good bolting practice takes training and is not necessarily intuitive.

LUBRICATION

Lubrication is the Rodney Dangerfield of the maintenance field. It gets no respect. People peripherally associated with maintenance assume that anyone who can find a zerk fitting and squeeze a handle can be a lubricator. Maintenance experts know that tribology is a rich field of study; you can even get a PhD in it. They also know that a good person in the lubricator's role can save a plant, building, or fleet thousands of dollars in breakdowns and potentially millions in downtime and accident prevention. The newspaper article in Exhibit 7-4 reports that a fire resulted from an overheated bearing. Note that the solution was to add sophisticated sensors — and not to upgrade the lubrication program.

Failure to lubricate results from several factors. A leading factor is poorly designed or installed equipment where either the lubrication points are too hard to reach or there are just too many of them. Other factors include too many different lubricants used, not enough time allowed, lack of standards, and a lack of motivation of the worker. The lack of motivation can usually be traced back to a lack of understanding how important lubrication is to reliability, poor self esteem resulting from a job which is regarded as the bottom of the barrel, and a lack of training and feedback on how the job was done.

Entry-level operators may take weeks to learn basic lubrication. We assume that journeymen mechanics are experts in lubrication. Yet frequently they know only what they've seen and tried. This might be only a small subset of the possibilities; their knowledge might also be outdated.

One issue is that many plants use too many different lubricants. In some cases you can standardize on the "better" product and save money through larger buys. The cost of the lubricant itself is usually the smallest element of the whole picture. If changes are made, document them and the reasons for the change. In most facilities, the lubricants were chosen long ago and the original reasons were lost in time.

Bangor News

Tuesday, July 28, 1992

Faulty bearing leads to fire on railroad car

NEW GLOUCESTER (AP) — Firefighters put out a stubborn fire Monday in a derailed boxcar that was packed to the ceiling with plywood, but only after the car was lifted back onto the tracks and pulled two miles to an area where the burning wood could be unloaded.

The pre-dawn mishap was traced to a faulty wheel bearing that became overheated, igniting the wood inside the boxcar and twisting the front set of wheels off the track along an isolated stretch in the woods near the Danville town line.

No one was injured, and none of the other 92 cars derailed.

The Auburn Fire Department delivered limited supplies of water with a special vehicle that traveled along the tracks. But firefighters could do little besides try to cool the outside of the boxcar because the tightly packed plywood prevented them from reaching the fire inside the front end.

"You might nickname it a hot box. It was cherry red, just like a stove," said Auburn Deputy Chief Robert DeWitt.

Railroad crews finally used a crane to lift the burning car and replace the wheel. Then it was hauled to Danville Junction, where the burning material was unloaded and the fire extinguished by late morning.

The derailment was reported at 1:45 a.m. About 30 firefighters were called in because the intense heat forced them to work in shifts, DeWitt said.

F. Colin Pease, executive vice president of Guilford Transportation Industries, said he was unsure how much of the plywood was destroyed but that "probably most of it" burned.

A company official was dispatched to inspect the damage to the car, Pease said in a telephone interview from his office in North Billerica, Mass.

The train was headed from Portland to Mattawamkeag. The car that derailed was 56th in the line of 93 cars, Pease said.

Pease said the design of the enclosed wheel bearings makes it impossible to examine them during routine inspections.

"It doesn't happen very often," he said.

He said the fact that the derailment was limited to only one car reaffirmed the wisdom of a company policy that limits train speeds to 40 mph. He speculated that this train was traveling at around 25 mph.

The company has installed a few "hot-box detectors" along its tracks south of Portland and is considering putting in more at other locations. The devices sense any excessive heat from passing trains and relay the information to the engineer or dispatcher.

Exhibit 7-4 Actual disaster could have been much worse.

For lubrication to be successful, the people involved need to understand why they are doing the lubrication, how to do it, where to do it, and with what. Drawings, charts, diagrams, and annotated photographs are useful in the process. The lubricator must also understand the implications of over-lubrication.

Cleaning and examining lubrication points are among the biggest areas where cleaning and lubrication overlap. Exhibit 7-5 is from Tim Gilles, an automotive teacher at Santa Barbara City College for over 35 years. He is also the author of several textbooks.

Clogged, dirty, or broken lubrication fittings compromise the whole effort. Initial cleaning should highlight these issues and correct them.

WALK-THROUGH AUDIT QUESTIONS

These questions were partially adapted from the TPM Development Program.

- Are lubricant containers always capped?
- Are the same containers used for the same lubricants every time?
- Are all containers properly labeled?
- Is the lubrication storage area clean?
- Are adequate stocks maintained?
- Is the stock area adequate in size, lighting, and handling equipment for the amount stored?
- Do you have an excellent long-term relationship with the lubricant vendor?
- Does the vendor's sales force know enough about tribology to solve problems?
- Do the vendors periodically tour the facility and make suggestions?

- Is there an adequate specification for frequency and amount of lubricant?
- Are there pictures on all equipment to show how, with what, and where to lubricate and clean?
- Are all zerk fittings, cups, and reservoirs filled, clean, and in good working order?
- Are all automated lubrication systems in good working order right now?
- Are all automated lubrication systems on PM task lists for cleaning, refilling, and inspection?
- Do you have evidence that the lubrication frequency and quantity is correct as specified (oil film onmoving parts, freedom from excess lubricant)?
- Is oil analysis used where appropriate?

CLEANING

Dirt is another enemy which is more lethal than most realize. Moisture in the wrong places and any other contamination are included in this conversation if they don't belong. Dirt can increase friction and heat, contaminate product, cause looseness from excessive wear, degrade the physical environment, cause potentially lethal electrical faults, contaminate whole processes (as in clean rooms), and demoralize the operators. Dirty equipment creates a negative attitude that adversely impacts overall care. Inspectors cannot see problems developing if they are hiding beneath a layer of dirt; mechanics don't want to work with equipment which has gotten too dirty.

Cleaning is a hands-on activity. Someone who cleans a machine with their eyes open will see all sorts of minor problems. If they are alert and tasked with it, they will also ask questions about how the

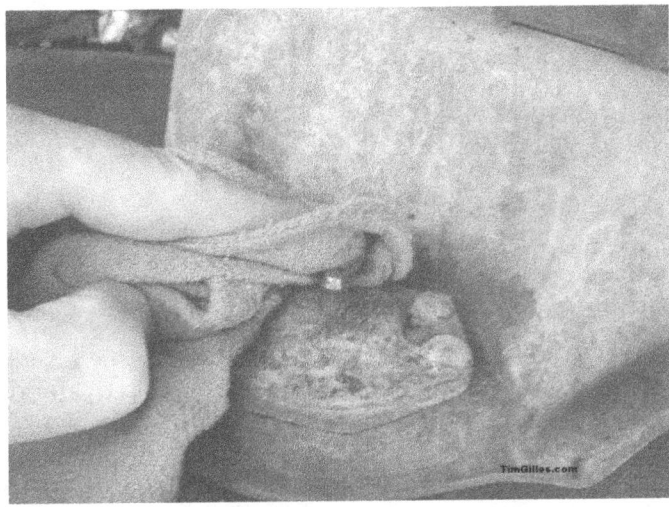

Exhibit 7-5 Picture showing a dirty Zerk fitting.

equipment works and why it is designed the way it is. Furthermore, they will increase their respect for the machine. This process of cleaning, seeing, touching, and respecting the machine is essential to increase reliability. As a result of the questions and observations made by people, cleaning the operation and maintenance of the machine can be improved.

Part of the cleaning process is to look for ways to make cleaning easier (called maintenance avoidance). Perhaps the source of dirt should be isolated to reduce the need for cleaning. In other cases, the machine could be moved or rotated to facilitate access.

TPM Development (see bibliography) lists seven steps to a cleaning program:

1. Cleaning the main body of the machine, checking and tightening bolts

2. Cleaning ancillary equipment, checking and tightening bolts
3. Cleaning lubrication areas before performing lubrication
4. Cleaning around equipment
5. Treating the causes of dirt, dust, leaks, andcontamination
6. Improving access to hard-to-reach areas
7. Developing cleaning standards

EXAMPLE: KEEP AREA CLEAN

This example looks at an automatic lube system needing a good cleaning (Exhibit 7-6). Keeping it clean is not only a PM issue. Cleanliness is important for rebuilds, major repairs, and even small repairs. Any mechanic in the business for a length of time can remember a perfect repair gone badly because of dirt.

Exhibit 7-6 Automated lubrication station needs a bit of cleaning. Otherwise what can you expect from this?

Exhibit 7-7 Precision rebuilding shop needs cleaning, painting, organization, and lighting. In short, a 5S project would be very valuable here.

With all of the attention being paid to dirt and cleaning, one would imagine organizations would take extra steps to exclude dirt when they do major repairs. Yet surprisingly few professional maintenance organizations take control of the physical environment with work tents, plastic drapes, or other measures to exclude dirt and contamination.

Keeping the maintenance shop clean should be a goal of the maintenance program. Cleaning the shop itself would come under the precepts of 5S. Issue a periodic work order to clean up the shop. Also look at eliminating the sources of dirt and clutter such as misplaced trash containers, lack of proper storage, broken tools, bad ventilation, inadequate lighting, and too small benches.

Exhibit 7-7 shows a precision engine / transmission, torque converter rebuilding shop for a surface mine. What are the chances of contamination and what are the impacts from that contamination?

How to Choose the Best Tasks

Start with a machine or area. Collect the O&M manual and any PM task lists already in existence. Look for tasks that are appropriate for operators or other factory workers without a lifelong commitment or experience in maintenance. Finally, concentrate on TLC tasks that are not being adequately managed now.

Tasks:
- Should be done daily or weekly
- Can be well defined
- Can be drawn or sketched with pictographic instructions
- Will give an excellent ROI in terms of uptime
- Do not require more than basic tools
- Do not expose the operator to excessive hazard

Descriptions Versus Specifications

In the beginning of my career, I worked for the U.S. Merchant Marine Service as a wiper. The wiper had the job to do all the dirty and nasty jobs on the ship. But it was a position where, if you wanted to move up, you could after your first six months of sea time. The next step up was the oiler.

I got my big break when the oiler got sick. They appointed me acting oiler. When the engineer showed me my new (albeit temporary) duties, he also indicated he expected I would be his eyes and ears into the health of the equipment. For example, he said that if anything here was unusually hot I should report it to him immediately. I just stood and said nothing; we were standing in an engine room where everything was unusually hot (as far as I was concerned).

Unusually hot is a description that requires experience and as such makes a poor specification. It is particularly poor for a newbie with about four weeks of sea time under his belt. What was needed was an instruction like "Report any machine in this area that is running above 250 C." And then give me some way to measure the temperature. *That* would be a clear specification and a clear way to measure it.

It may sound strange, but one thing we in TPM want to get away from is requiring judgment and experience for decision making. The TPM operator needs rules (like "report to me any machine operating above 250 C°) and a way to measure or inspect the asset to see if the rule is broken or intact. Of course, judgment and experience are great, but they take too long to develop and are too variable from person to person.

Team Meetings

Team meetings are important to the success of TPM. Good meeting discipline is important so that the meetings do not devolve into bull sessions or blame storming sessions.

One of the most important times is the first meeting. If the attendees are not regular meeting goers, it is essential to get off on the right foot. According to the Penn State site for meetings, the goals of the first meeting are usually to "reaffirm" the project goals, and then establish ground rules and communication lines for how the team should operate and what the intended goals are. In our language, we use the first meeting to remind people about the goals for TPM and why we are doing it.

Bring the TPM charter or other project assignment documents — being able to refer to original documents during the meeting will help you clarify your planning.

- *Name your team.* This may sound very trivial but having a common name is a good way to feel closer to the project. It is surprisingly important to pick a cool name!
- *Share contact information.* You will probably want to share e-mail addresses, and possibly phone numbers. You should also establish when and how different tools should be used.
- *Establish a timeline, roles, and responsibilities.*
- *Set ground rules.* Although disagreements will arise, it is possible to voice opinions in such a way to that conflicts do not escalate. Typically, it is suggested that personal attacks be avoided. See below for additional ground rules.

One of the most important ideas is to have a specific item or problem to discuss or solve. When you call meetings, do the following:
- Circulate the date, time, and location of the meeting. Sending reminders the day before the meeting may be wise in some cases.
- List the attendees expected.
- State the purpose of the meeting.
- Detail the order of business to be conducted at the meeting.

- Describe where to find background or support materials required.

The more focused the meeting is, the better. Meetings are time killers if they are not managed. It is important to have ground rules for these meetings:

- When you start up TPM meetings have everyone review these (or your own) rules.
- Be prepared to drop a topic if it gets bogged down and think of another approach.
- Have notes or minutes from prior meetings.
- Run them on time and insist everyone attend on time.
- Encourage everyone to participate in the meeting.
- Everybody comes prepared.
- Manage multiple conversations, off-board discussion (discussions on another topic) and comments; it's very disruptive having another discussion going on.
- Don't text, tweet, answer E-mail, or take phone calls in the room unless it is related to the topic at   hand.
- Pay attention to everyone speaking; have an open mind.
- Be patient and calm.
- Be sure to complete any tasks assigned.
- Summarize decisions and future plans before you leave the meeting, especially action items.

Meeting Quiz

If you want to check out your meeting muscle, take the following quiz (developed at Penn State University to help them evaluate their meetings).

Evaluate the meetings with the intent of improving future meetings. Ask yourselves the following as a team (or you as a member).
- Was the purpose of the meeting clear?
- Did the set up of the room help or hinder the meeting process?
- Could the room set up, logistics, or location be improved for meetings?
- Was jumping to conclusions allowed?
- Did the group help to suspend judgment and explore alternatives?
- Did the group use conflict in a positive way to differentiate ideas?
- Did the group work toward consensus?
- Did the team leader document the interaction when the process seemed ineffective?
- Did the group insist on action commitments (what is to be done, by when, and by whom)?
- Did the group identify a follow-up processes?

Typical Team Activity: Conducting a 1-Point Lesson

WHY DO A 1-POINT LESSON

The 1-point lesson communicates knowledge and skill about the asset among members of the team. The communication makes sure that everyone knows about a better way of doing something. Then, the next time a problem is encountered, everyone knows the way to solve it.

The 1-point lesson raises the knowledge and skills of the team in a very short period of time. It also can be used for eliminating problems and making improvements to the way of working.

HOW IS IT DONE?

One member of the team prepares a sheet that describes the problem and solution in simple language. It is often illustrated with pictures if possible.

- The team discusses the 1-point lesson, formally or informally, and incorporates any ideas.
- The 1-point lesson is approved by the line management to ensure it is appropriate and safe.
- The 1-point lesson is then published and read by all members of all shift teams.
- Individuals are asked to sign an acknowledgement that they have read the 1-point lesson.

This is very similar to a Root Cause Analysis (RCA), but simpler; it might be called RCA Lite. One-point lesson techniques apply to only the simplest problems. For problems with any more complexity and resistance to solution, RCA is the right tool. In fact, the TPM team can decide if enough is known to use a one-point lesson or an RCA. The follow up and bedding down of the improvement are the same.

KEYS TO A SUCCESSFUL 1-POINT LESSON

- Treat only one piece of knowledge at a time.
- Use simple clear language.
- Illustrate the idea with pictures or drawings.
- Make the 1-point lesson available to everyone.
- Recognize and reward those who share their knowledge.

Here are some simple examples:
- Match marks painted on bolts and fittings
- Improved procedures for size change

- Procedures for efficient recovery from a machine stoppage
- Instructions for the repair of a breakdown
- Point out the optimum settings for a machine or process

What should the TPM 1-point lesson contain? Each element of the TPM tasking can be developed through a 1-point lesson.

- Before and after sections (the use of photographs or diagrams is encouraged)
- Plant description and location
- A description of the situation or the solution
- Who prepared and who approved the document?
- Date of introduction
- Information and conclusions which are is easily understood.
- A suitable system embedded within it to ensure compliance and acceptance.

The one-point lesson information is courtesy of Keith Rimmer, a senior consultant who has worked worldwide on Lean and TPM projects.

A great archive of single point lessons can be found at the web site of Fuss & O'Neill. They are a full-service engineering consulting firm: http://www.fando.com/News_&_Resources/Single_Point_Lessons/SPL_Archive/

Prioritization of Losses

Okay, you have started a road toward TPM and autonomous maintenance. As mentioned, the first step is to get up close to the machine and clean it. You will start to develop a list of problem areas. TPM teams need to triage the issues they face. My advice is to go for the easiest problems first. Pick the low hanging fruit.

If the team doesn't prioritize the problems they find, they will get bogged down, which will demoralize the team. There are two major categories of problems. Problems with a line or machine can be looked at as problems which come up sporadically; these can be addressed by the TPM team. Chronic problems need specialized knowledge, skills, and aptitudes.

When a problem goes beyond the expertise of the TPM operators, they have to let it go to maintenance and engineering without spending too much effort. Exhibit 7-8 distinguishes the differences between the two types of losses.

RCA (Root Cause Analysis) for Operators

One of the most powerful activities of the TPM team is root cause analysis (RCA). It is completely appropriate for operators to work on RCA projects. The most effective RCA teams have a diversity of members, but at least someone close to the problem from operations, maintenance, and engineering. Other parties are also appreciated as long as they can bring another point of view to the problem. The team size should be 3-to-6 people.

Root cause analysis is a process that follows causes of problems. It is called "Root Cause" because it is supposed to track the causes back to find the source or "root." Unfortunately, according to Bernie Piovesan (a leading trainer in RCA for RCA RT (Round Table) in Australia), "Root cause Analysis almost never finds 1 root cause because usually there are several interrelated causes of any major problems." This is especially true for the chronic problems described

Aspect	Sporadic Loss	Chronic Loss
Causation	Causes for this failure are easy to trace and cause-effect is simple to understand	This loss cannot be easily identified and solved even if countermeasures are applied
Remedy	Easy to establish	The cause is hidden defects in the machine, process, part or method
Impact/Loss	A single loss can be costly	A single cause is rare; usually a combination of causes
Frequency of occurrence	Frequency of cause is low to occasional	Frequency is higher
Types of analysis	5-Why	Intricate and complex methods required, including cause and effect, correlation
Corrective action	Line personnel and operators	Specialists in engineering, quality, and maintenance

Exhibit 7-8 The difference between sporadic and chronic losses

above. He goes on to say we probably should call it "cause analysis," for that is what we actually do.

There are several different systems of RCA. In fact, Bill Holmes (also of RCA RT) says that successful RCA covers a spectrum of activities and approaches. For example, the RCA process for a small production issue would vary in rigor and even approach to an investigation of a fatality.

Probably the most popular approach is called 5 Whys. With this approach, someone on the team asks the question "Why" 5 times. Five is not a magical number; it just seems to be enough iteration to get below the surface to one of the root causes.

In more comprehensive RCA, there are three parts to the team's efforts.

Define the problem. Defining the problem is an essential step. It is expressed as a combination of something that is not happening, a description of what should be happening, and a measure of the gap between the two.

Look for causes. In this step, the team looks at the potential causes. Many techniques can be used to come up with causes. The important thing here is to look far and wide for causes. They may be in one or more of several domains, including engineering, operations, maintenance, design, weather, procedure, and many others.

This part of the analysis can be started by a brainstorming session. People just list all the causes they can think of. In brainstorming even weird, stupid, or ignorant causes are recorded. The reason is that a stupid idea (aliens hit it with a ray) might stimulate someone to think of a great idea (rays of sunlight heated it and caused an expansion).

The process divides the causes into primary (leading directly to the problem) and secondary (leading to a cause leading to the problem. It is important to distinguish the primary from the secondary causes. After a bunch of causes are identified, the team starts to build what is called a cause tree. New causes may come to light and can be added at this point.

The cause tree tracks from the problem through the primary causes, to the secondary and tertiary causes, theoretically back to the root. You follow the path backwards until you start to leave the scope and authority of the team. To follow an individual branch backwards the "5 Why" technique is useful.

Fix forever. In this final step, we want to go beyond identification of the causes to interrupting the cause tree so the problem does not occur. The change has to be tested and thought through so there are no unintended consequences. Finally, the fix has to be documented and communicated so it becomes the norm. The solution should be followed up to see if the fix stuck and continues to fix the problem (without causing new problems). Information and a free web based RCA system can be found at www.RCART.com.au.

Lean Maintenance: Is TPM Lean? TPM is Lean!

Lean Maintenance is a sibling to TPM. They are both children of TPS. The 6 (or more) losses to production can be seen as 6 wastes. Any activity cutting waste is Lean.

TPM is an important part of traditional Lean Manufacturing. From a Lean point of view, TPM offers three opportunities. One is time savings, because the operator is usually already in close proximity to the machine, so travel time is reduced. The operator is also already in control of the machine, so custody transfer is simplified.

The second opportunity is that TPM tasks feature lower skills (lower than full-blown maintenance activity) and is also usually a skill upgrade for the operator rather than a skill downgrade for maintenance professionals.

Finally the goal of TPM is high OEE which is also the goal of Lean activity (although it is not called that).

The advantages of TPM flow from higher productivity and higher quality output, and also from the effects of the improvements in attitude. The higher productivity (which is Lean) comes from having the operator do basic maintenance, improve conditions, and eliminate troubles when they are small.

The operators are already at the asset, already have custody, and already have a job assignment. No additional travel is needed for the operator. The improvements in attitude come from the shift from a passive stance as an operator toward an active stance, becoming the person fully accountable for the whole thing, and an expert in machine health.

Most TPM tasks are also PM tasks when they are being done by maintenance personnel. The Leanness comes from having the right person do the right tasks, and the reduction of non-productive time (such as travel). Moving the maintenance effort to the section of operations that is located on or near the asset will be Lean. We have to remember that too much TPM is the same as too much PM. It is Fat.

TPM naturally attacks the sources of ineffectiveness. Most companies spend enormous amounts of money on improvements in efficiency. Efficiency is defined as doing things the right way or getting the most output for the least input. TPM can be said to take the next step. TPM looks at doing the right things right. By attacking all the losses that impact production, TPM ensures that, at the end of the day, the pile of good salable parts made by the process is bigger. Sometimes, after a TPM implementation, the pile of good parts is a lot bigger.

TPM gives significantly increased power to the operators. TPM works only because the operators begin to own the equipment. As ownership spreads, autonomous maintenance becomes a reality. TPM works inside the culture of the organization, to transform the relationships of people to the output. This new power changes people's minds about the areas in which they can have an impact.

Chapter 8
Facts of Life

What Are We Trying to Do?

The TPM team looks to eliminate all reasons for downtime and reduced output. In a maintenance-oriented class, we concentrate on the maintenance tasks. In an operations class, we focus on operational issues. However, a full TPM effort would look at the data for all the losses and work on each one.

For breakdowns, TPM tasks should be directed at how the asset could fail. The rule is the tasks should eliminate or notify us of the:

- Most expensive…
- Most likely …   failure modes
- Most dangerous…

To go after breakdowns (remember only 1 of the 6 losses), we have to figure out where they come from. Breakdowns can be rooted in maintenance, tooling, operational processes, errors, materials, and engineering of the part, product, or asset.

To solve the breakdown problems, tasks are assembled into task lists, changes are made to processes, machines are mistake proofed, and engineering is reviewed. The task list and new operational processes might be printed on a placard attached to the machine (visual maintenance).

Caveat: Failures and breakdowns will occur even with the best PM and TPM systems in place. Your goal is to reduce the breakdowns to minuscule levels and convert the breakdowns that are left into learning experiences to improve your delivery of maintenance service.

One Problem However — Past Sins

TPM/PM systems fail because of past sins (accumulated deterioration). These sins wreak havoc on any task force trying to change from a fire fighting operation to a PM or TPM operation.

The first step is often to bring the asset up to full operational spec-

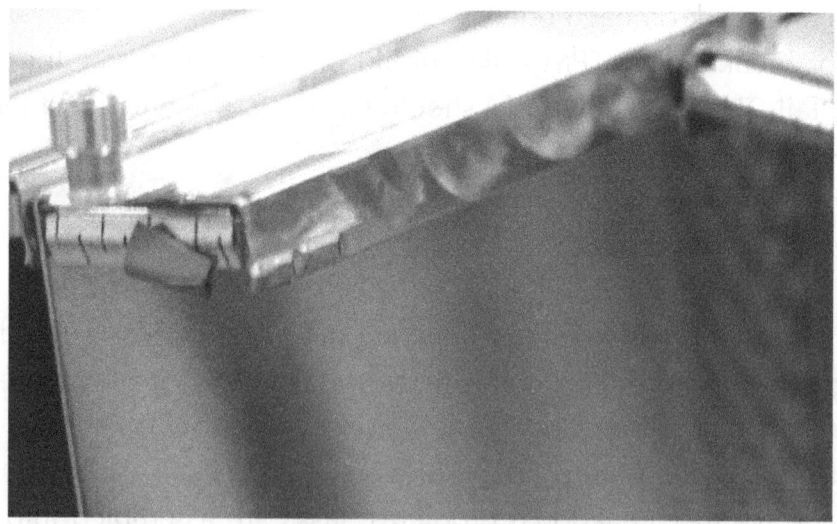

Exhibit 8-1 Damaged shielding would have to be replaced and returned to like new Condition

ifications. Return it to "like new" conditions. As you do that, you have to face unfunded maintenance liabilities and fund them by doing the repairs. The only way through this jungle is to pay the piper, modernize, and rebuild out of the woods. Any sale of a TPM system to top management must include a budget line item for investing in bringing the asset base up the standards — rehabilitation, rebuilding, renewal, or replacement. Remember, wealth was removed from the equipment every year it went without sufficient maintenance funds invested to keep it in top operating condition.

As mentioned, one essential element for the TPM team is to communicate with the maintenance department about any issues with the equipment beyond the scope of TPM. TPM operators are responsible for writing up any conditions requiring attention, and then turning their reports over to maintenance in a timely manner. The work order system part of the CMMS is the method in most plants for formal communications with maintenance. When TPM initially starts up, maintenance should have specifically allocated resources to deal with the past sins.

Three Life Cycle Phases of Equipment

TPM is designed to closely follow the equipment's condition. Therefore, different task lists, approaches, and attitudes are needed to cope with machinery in various conditions.

1. The first life cycle phase is when you buy, build, and install the equipment. Here we call it the start-up phase. It is also called the infantile mortality phase.
2. The second phase is running well and making money. Here we call it the wealth phase. This is the goal for the TPM effort.

3. The final phase is the breakdown phase. TPM is designed to never allow an asset to reach the breakdown phase.

Each phase has characteristic problems and solutions. The TPM actions will be different. It is important to know the phase for each piece of equipment (in some cases each component of the equipment is in a different phase) so it can be addressed appropriately.

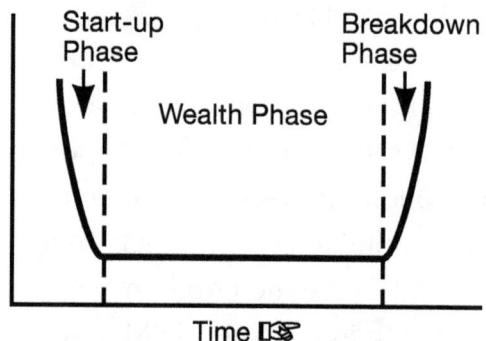

Exhibit 8-2   The Bathtub Curve

In the bathtub curve in Exhibit 8-2, the number of incidents of failure is represented by the vertical axis and time or equipment use is the horizontal axis. This illustration shows a lot of problems on start-up, decreasing to the point where you start to "make money" with the asset. The wealth phase lasts until the asset has had enough wear and tear that large components are starting to fail.

We will now review the phases in detail, including the kind of interventions the TPM team can make in each.

1. Start-up Phase

Equipment tends to have many more problems when newly installed. Assuming it survives to lead a healthy and productive life, we can characterize this beginning phase as the infant mortality phase or, graphically, the beginning of the so called "bathtub curve." Failures due to defects in materials, workmanship, installation, or operator training on new equipment show up quickly when first brought in; later problems tend to be less severe or less frequent. Usually these early-stage costs are partially covered by warranty. Unless you have significant experience with the make and model of the asset, you will be hampered by a lack of historical data. The failures are hard to predict or plan, and it is difficult to know which parts to stock.

This period could last from a day or less to several years on a complex system. A new punch press might take a few weeks to get through this cycle, whereas an automobile assembly line might take 12 months or more to completely shake down. Be vigilant in monitoring misapplications (the wrong machine for the job), inadequate engineering, and manufacturer deficiencies.

Countermeasures to shepherd the asset through the problematic start-up cycle include:

- Enough time to test-run equipment
- Enough time to design and build the equipment if it is custom
- Enough time and resources to install properly and then commission
- Operator training and participation in start-up
- Operator certification
- Operator and maintenance department input into choosing machine

- Maintenance and operator inputs to machine design to insure maintainability
- Good vendor relations so they will communicate problems other users have
- Good vendor relations so you will be introduced to the engineers behind the scenes
- Training with the equipment (and periodic retraining as the project continues)
- Maintenance person training in start-up
- Latent defect analyses (run the machine over-speed to see what fails, then re-engineer it)
- RCM analysis to design PM tasks and reengineering tasks
- Rebuild or re-engineer to your own higher standard
- Formal procedures for start-up and commissioning

Operators that have been through this stage can be of great help by being trained and then by being in the initial stages of TPM (close-in cleaning, tightening, and correction of minor faults).

2. Wealth Phase

All curves have a wealth phase, unless the asset is not strong enough for the job; those go directly from start-up problems to the breakdown cycle. The wealth part of the cycle is where the organization makes its money from the useful output of the machine, building, or other asset. This can also be called the use cycle. The goal of PM is to keep the equipment in this cycle as long as possible, or detect when it seems to be making the transition to the breakdown cycle. After detecting a problem with the machine or asset, a quality-oriented maintenance shop will do everything possible to repair the problem and keep it generating wealth.

After proper start-up, the failures during this cycle should be minimal. Operator mistakes, sabotage, and material defects are particularly obvious during this cycle if the PM system is effective. Remember that PM would generate the need for Planned Component Replacement (PCR). The wealth cycle can last from several years to 100 years or more on certain types of equipment. The wealth cycle on a high-speed press might be 5 years whereas the same cycle might span 50 years for a low speed punch press in light service.

Countermeasures designed to keep the asset in the wealth cycle include:

- PM systems
- TLC (tighten, lubricate, clean)
- Operator certification
- Periodic operator refresher courses
- Close watching during labor strife
- Audit maintenance procedures and check assumptions on a periodic basis
- Audit operations procedures and check assumptions on a periodic basis
- Autonomous maintenance standards
- Quality audits
- Quality control charts initiate maintenance service when control limits cannot be held
- Membership in user or trade groups concerned with this asset

The goal of all of the maintenance effort is to keep the equipment in this phase to maximize the profit for the company.

3. Breakdown Phase

This is the cycle that organizations find themselves in, all too soon, when they do not follow good PM practices. The breakdown phase is characterized by wear-out failures, breakdowns, corrosion failures, fatigue, downtime, and general headaches.

From a maintenance viewpoint, the breakdown phase is a very exciting environment because you never know what is going to break, blow-out, smash up, or cause general mayhem. Some organizations manage the breakdown cycle very well and save money by having extra machines, low quality requirements, and toleration for headaches. Parts usage also changes as you move more deeply into life cycle three. The parts tend to be bigger, more expensive, and harder to get.

The goal of most maintenance operations is to identify when an asset is slipping into the breakdown phase and fix the problem. Fixing the problem will result in the asset moving back to the wealth phase or, if it has had a major rehabilitation, back to the start-up phase.

Countermeasures for the breakdown cycle include:
- PM systems (particularly TLC)
- Maintenance improvements
- Equipment re-engineering
- Reliability engineering
- Maintenance engineering
- Feedback failure history to PM task lists
- Great fire fighting capabilities
- Superior major repair capabilities
- Great relationships with contractors who have superior repair and rebuild capabilities

Equipment cannot be run under TPM while it is in near the break-down phase. That is a bad neighborhood and will discourage the operators and eventually render the TPM system ineffective.

Chapter 9
CMMS
(Computerized Maintenance Management System)

TPM was implemented in Japan before the widespread use of computer systems for managing maintenance in that country. The issue was never dealt with in the original literature, training, and consultation assignments. Subsequent books on TPM (since the 1990s) seem also to sidestep the computerization issue.

If we delve into the reasons for this, we might see it is related to the dislike of everyone for additional paperwork. Maintenance professionals who are intent on getting operations to take over basic PM want to sweeten the pot by promising "no paperwork" as if extra paperwork would sour the deal.

Does this make sense? Perhaps having TPM activity without feedback to the CMMS does make sense. It would save time and effort. It would lower training requirements of the operators. Every operator would tell you that the production control system and the quality systems already make them do too much paperwork.

The CMMS is the center of data about the maintenance and reliability of the assets. Systems with effective (that is complete) databases have a complete set of information related to the costs, frequencies, tasks, time spent, and materials to keep an asset running. In short, some of the benefit comes from having all the information in one place. Of course, it has been a struggle to accomplish this.

Having the operators generate maintenance incidents (PM, corrective or otherwise) means they are for a short time part of the maintenance department. As part of the department, it is important for the TPM effort to add their data to the overall database. The challenge is to have the data available for analysis without excessive cost, paperwork, or time. Also, the sheer number of TPM transactions would present an analysis barrier.

MRP II Interface

The system in control of the machinery (production control) is the MRP II system (Materials Requirements Planning II — modern version). To MRP, the TPM activity looks like set up or other supportive activity. The time for the TPM activity must be programmed into the MRP system so that the lost production time is accounted for.

In the best possible world, the MRP system would generate a standing work order for a week or month's worth of TPM activity. Or the MRP system would close the PM work order for the TPM activity with hours spent and materials consumed. The TPM operators would have their maintenance time logged to the CMMS via the MRP system.

According to Mark Goldstein, Ph.D., the MRP system is very much like a maintenance scheduling system in a general sense; thus, there are analogues of the various data flows. In MRP, the scheduling function attempts to bring together the materials (all resources based on the bill of material), the machine (custody of the asset for production), and the appropriately trained person. The MRP schedule includes a minute-by-minute schedule for the machine. It allows for a certain run time to produce the forecasted number of pieces, set-up time (again person, tooling, and machine custody), model change, and size change times and should allow for scheduled PM (and TPM) activity. If we needed access to the asset, the MRP system would be the logical system to go. It is also the best system to feed the CMMS.

The specifics of TPM are another story. The task lists would be stored in the CMMS. The TPM operators would use the structures of the CMMS to store instructions and would use any external systems that maintenance personnel use for task justification, task engineering, and drafts of task lists.

TPM's goal is to involve the operators and make them accountable for the asset's health and, ultimately, reliability to the extent of their mission or scope of work. Where TPM leaves off, the maintenance department picks up. The goal of maintenance is machine capacity. The goals support each other.

The CMMS Work Order

The interface between the TPM operator and the maintenance system is the work order (Exhibit 9-1). In some systems, the work order is known as the work request or notification. Training is going to be needed in a few key areas.

If the repair is done by the TPM operators, then they would be expected to fill out the entire work order, including the sections noted "by mechanic."

Review this field	For this information	Notes
Work requested	Specific, observable, actionable	This is the single most important field with a corrective order. Requestors usually put their conclusion down (pump broke) and their solution (replace pump). Although this might be okay some of the time, a more useful approach is for the requestor to put the specific observable phenomenon, such as 0 gauge pressure on outbound side of pump 3S on the Stoker line. They might add a comment (e.g., "I think the pump is broke").
Asset #	Correct, readable, same name as in CMMS?	It is important that everyone use the same language and referral names. If necessary, paint the name on the side of the machine or, if possible, put a cross reference chart into the CMMS.
Reason for repair	Is the category correct?	The reason for the repair is an important field for future analysis. Any kind of improvement project should be distinguished from regular maintenance. Any maintenance done as the result of an inspection activity should be coded (as CM) differently than jobs found by breakdown.
Priority	Does the priority make sense given the job requested?	There should be an established, well-understood priority system. The request should conform to this system.
Timing (when needed)	Is the date needed reasonable given the request?	Operations supervisor or TPM operator
Location	Is the location where the work to be done clear?	Should be noted by the TPM operator if there is any question
Proper account to charge if needed	Is the account correct, is it filled in if necessary?	Should be well known for all work
What was found?	Concise information about what caused the work requested.	By mechanic but should be the same as the work requested
What was actually done?	What was done to fix or mitigate the problem found? Should be short and follow plant conventions	By mechanic, in some cases the TPM operator assists
Materials actually used	Correctly identify materials and add any free issue items.	By mechanic from warehouse tied to work order number
Special tools and equipment used	Identify special tools, equipment, and scaffolding.	By mechanic
Outside services used	Note any outside services used on this work order	By planner, supervisor, stock room, or purchasing
Time in-out 1	Elapsed time must be accurate to +15 minutes.	By mechanic
Time in-out 2	As far as practical, correct the job steps and update estimates for each step	By mechanic
Time in-out 3	If downtime is	By mechanic

Exhibit 9-1 Some of the fields of a CMMS Work Order

Work Order Training Program for TPM Operator

. The key to effective future analysis is in the information from the TPM Operator. The important question asks how many people realize how important their notes and entries are on the work order. One effective teaching method is to ask the TPM team members to participate in some basic analysis from the existing data. They might find that analysis difficult because of problems with the data integrity. This course can be an outgrowth of those difficulties.

The very simplest questions cannot be answered without cooperation — questions like how much did we spend maintaining this machine? There are three issues:

1. The completeness of the data entry
2. The accuracy of the data entry
3. The consistency of the data entry nomenclature (the same thing must be called the same thing each time)

The work request should answer the following questions posed by Jay Butler in Maintenance Management (seminar work book used at Clemson University's Maintenance Management class):

- Is there a possible safety or environmental implication to this failure?
- Where should the maintenance person go?
- What is wrong? What was observed?
- What unit is to be worked upon?
- Who should they see when they get to the area?
- What was happening just before failure?
- What does the caller think is the problem?
- How critical is the unit or process?
- What time and date did the call came in?

Work requested	This is the field that usually needs the most training. TPM Operators tend to put down their conclusion (pump broke) and their solution (replace pump). Although this might be okay some of the time, a more useful approach is for the operator to put the specific observable phenomenon such as 0 gauge pressures on outbound side of pump 3S on the Stoker line. They might add in a comment (e.g., "I think the pump is broke").
Asset #	In some cases, the shop floor operators have names for the machines different than the maintenance department files. It is important that everyone use the same language and referral names. If necessary, paint the name on the side of the machine or, if possible, put a cross reference chart into the CMMS.
Reason for repair	The reason for the repair is an important field for future analysis. Any kind of improvement project should be distinguished from regular maintenance. Any maintenance done as the result of an inspection activity should be coded differently than jobs found by breakdown.
Priority	There should be an established, well-understood priority system. The request should conform to this system. If your system does not allow the requestor to set even preliminary priority, then skip this field.
Materials actually used	The job plan might call out certain materials. The goal is to get accurate materials and fix any problems in the planned job package (if any).The actual job might use some, none, or all of the materials. Additional materials would be added with notes about if this was special extra work or materials that should be added to the work plan. Supplies and free access materials should be recorded also.
What was found	What did the mechanic determine was wrong with the system, if anything? An accurate statement can help future diagnosis because the service requested can be linked to what was actually found. In the future, if that same service is requested, the next mechanic may be advised to look at the same cause — in addition to other causes. This can also be used for MTBF analysis (component life, for example) so a consistent description or code should be used.

Exhibit 9-2   continued on the next page

Work actually completed	Based on what was found what work was accomplished. Some systems have work accomplished (WA) codes, which simplifies this entry. Some WA codes might be R&R (remove and replace), Repair, Rebuild, Adjust
Special Tools and equipment	The goal is to correct the planned job package with what was actually needed for the job.
Other Resources used	This would be the place where the mechanic could correct the plan as to lifting equipment, lifting gear, special tools not included above, etc.
Outside services used	Note any outside services used on this work order. This would include the use of contractors, inspectors, NDT, and engineers, along with what they did.
Time	The total hours spent on the job is essential for analysis. Agreement about how the time for breaks, interruptions, and lunch are handled has to be made.
Time	One goal from the training is to get input from the trades people to correct the job plan (if you operate in a planned environment).
Time	Maintenance downtime calculations (time that the service request came in – time returned to operations) are based on accurate entry of when the unit was returned to operations from the body of the work order.
Notes and comments	Anything that the mechanic thought was important may be very important at a later date. This might include diagrams and sketches, specialized information and problems with the job plan, or problems with the permitting, in short, anything that the mechanic wants to pass on to the next person.

Exhibit 9-2 Subset of the Work Order for TPM

The Simplified Work Order

One of the best solutions to the work order dilemma is to create a subset work order (Exhibit 9-2 on the previous page). Some organizations build a simplified front end onto their CMMS that has limited options for the work order. This can speed up the data entry and, if properly designed, can get better information at less cost. Automation using smart phones, PDAs, net books, or laptops can also be used to enter data.

Chapter 10
Visual Work Place

Gantt and His Charts

The visual workplace probably started with Henry L Gantt. In 1914 he was working at the Frankford Arsenal in Philadelphia. Mr. Gantt was trying to develop a system to show the expected artillery shell production versus the actual production. His challenge was that the workers were not universally literate and many were not strong English speakers. He developed the Gantt chart, the horizontal bar chart for which he became famous. He survived until 1919 and continued to work on displays of production statistics that could be understood with minimal English.

What he found is that people can read and get the meaning from a display or chart faster and more accurately than they could from a written report. This approach is the underpinning of the visual workplace.

Visual management is an extension of Gantt's vision. Visual tools — signs, labels, colors, charts, markings, signals, strips, tabs, handcrafts, colored edges. etc. — are used to control, communicate about, and simplify work processes. By using visual techniques to convey information, it is easily accessible to those who need it and the current status of processes is immediately apparent. This allows the organization to streamline work processes and eliminate waste.

Below are a series of examples of visual aids widely used in TPM and PM applications. Visuals enhance equipment reliability and maintenance efficiency. Lean manufacturers know that reliable production processes are critical to their operations. Reliable processes require immediate access to certain information and the visual workplace provides instant access to that information.

Benefits

Incorporating visuals into your reliability program can provide important benefits:

- Simplified operator-based care training (with cue cards all around to remind them)
- More consistency of tasks
- Faster detection of operating abnormalities (can compare with pictures on machine)
- Quicker troubleshooting and root cause analysis (charts in visual range)
- Reduced stores inventory (fewer mistakes)
- Fewer unplanned MRO purchases (better defined products and fewer mistakes)
- Improved safety and employee morale (knowledge provided finstantly is power and power makes a worker feel motivated)

Examples

Brady is a leader in providing the tools for the visual work place. Some of the items in Brady's line for TPM and PM activity are reproduced in Exhibits 10-1 through 10-5.

http://www.bradyid.com/bradyid/cms/contentView.do/0/3334/54 16/0/VisualWorkplace/Visuals-Enhance-Equipment-Reliability-and-Maintenance-Efficiency.html

Exhibit 10-1 Lubrication Visuals "Label created using the Brady GlobalMark(R)

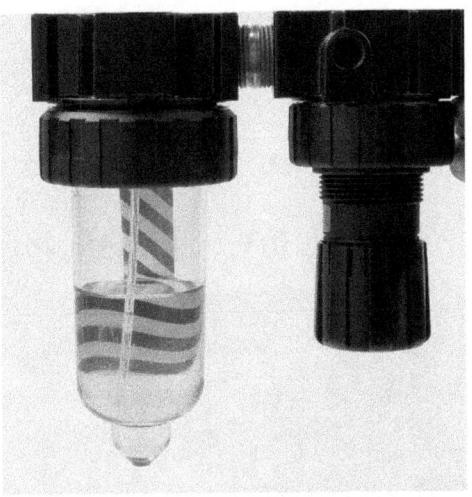

Lubrication errors (both under and over lubrication) are one of the major causes of equipment failures. Ensure employees can easily find the lube point, readily detect the proper levels, and correctly apply the right amount of lubricant at the right time.

Exhibit 10-2 Inspection Visuals

Make it easy for operators or anyone in the plant to quickly detect operating abnormalities and emerging failures before they become a problem.

Exhibit 10-3 Speed Troubleshooting and Repair

Including "to" and "from" information on equipment ID labels makes it easier to trace lines in electrical and piping systems.

Exhibit 10-4 Storeroom Management

 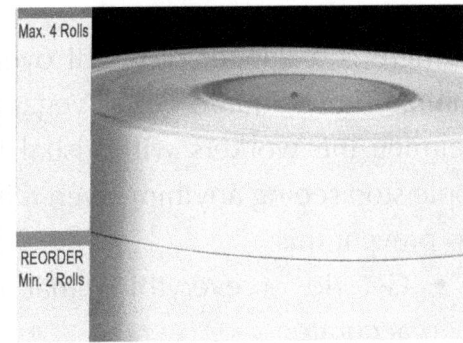

Up to 25% of downtime is taken up with parts-related issues — going to get parts, looking for the right part, ordering parts, etc. Visuals can greatly simplify these tasks, resulting in faster turnaround and lower inventory costs.

Exhibit 10-5 Predictive Maintenance

Use visuals to ensure that vibration, ultrasound, and infrared measurements are taken at the exact same spot each time — no matter who takes the reading — to ensure comparable data over time.

When Is Too Much Too Much?

There is a danger of visual overload. To help imagine overload, visualize Times Square or Las Vegas at night. You run the risk of overwhelming the workers with visual information. Once that happens, people stop seeing anything even if it is right in front of their noses! To help prevent this:

- Get rid of everything that is not current, up to date, and accurate.
- Get rid of everything that is not directly relevant to the job at hand and move it outside the workplace area (such as HR posters or "motivational" posters).
- Have the TPM teams review all visual displays and reduce, eliminate, or move the material around to be more effective.
- Apply 5S to the visual work place.

Visual management as originated by Gantt is a powerful tool for communicating to the employee both to be aware of managerial targets and to help them contribute to the achievement of the targets. The visual workplace provides information for workers to reinforce their own performance through displays of targets reached and progress made towards goals.

For the TPM worker, the visual workplace has a more immediate impact. When they do their TPM task lists, they refer to the pictographs on the machine, the signs in the work area, and the charts hanging on clipboards in their workplace.

Chapter 11
TPM Installation

Successful TPM Installations

What are we trying to do? What exactly are we trying to install?

Because we have learned that TPM is not a sly way to offload maintenance work on the production/operations personnel, we have an implementation problem. Changing culture is one thing. But we are trying to change people's long-held beliefs about who can do what on the shop floor. In short, we are attempting to wake people up and have them spontaneously and autonomously take care of complicated equipment. Not only that, but we are also trying to convince everyone that this group — which heretofore was to blame for all ills (at least in the eyes of the maintenance folks) — can think, reduce waste, solve problems, and improve the equipment.

That is a tall order. And that kind of transformation requires a two-prong approach. On one prong, we have to deal with what is on the surface. The second prong is dealing with the conversations below the surface.

Successful TPM installations require spending a good deal of time preparing for and training all the people involved. This training is an excellent way of dealing with all the surface conversations. This involves addressing four very basic requirements: motivation, long-term support, competency, and environment. These requirements prepare the soil for the transplantation of TPM concepts and attitudes:

Motivation

The entire staff and all the workers need to be open to a change of attitude toward all types of waste. Running Lean Maintenance projects is a way to jump start the process and is a great barrier jumper. This acceptance of the participation, energy, and intelligence of operators on the teams takes place over a long period of time as organizations run projects. In turn, projects soften the attitudes because when they are done, they go away; they provide less of the threat to the status quo.

Long Term Support

Both long-term support and toleration of bumps in the program implementation are needed from top management because TPM is a long-term cultural change. Motivation will happen when the workers feel management has their back and will stick with the program.

Competency

Certain skills are necessary before TPM can succeed. Every group has to take on new duties and will need new skills. Training operators in PM, design engineers, operators, and mechanics in root failure analysis will eliminate waste and losses. One assumption of TPM is that operators have capabilities that are not being used in their current job. With training, these capabilities are gradually utilized. Management must tolerate experiments, mistakes, and learning that accompany the experience.

Environment

Top managers in the organization must do more than support the thrust for improvement. They must be willing to lead the change of their own stereotypes and prejudices. The top people must understand the need for TPM and its implementation. Because this is a longer-term project, the operating environment will shift. That shift has to be husbanded by top management working with shop floor leadership. TPM goals should be included into the reports and meeting reporting.

Activities for Introducing TPM in an Organization

(Adapted from J. Venkatesh see article at http://www.plant-maintenance.com/articles/tpm_intro.pdf)

Announcement by management about TPM in the organization: Proper understanding, commitment, and active involvement of the top management are necessary, especially in the early phases of the program. Senior management should promote awareness programs after the announcement is made to all. Publish it in the house magazine and put it in the notice board. Send a letter to all concerned individuals.

- *Initial education and propaganda for TPM*: Training should be done based on the need. Sometimes intensive training is called for; sometimes just raising the awareness is enough. Take decision makers to places where TPM is already successfully imple mented.
- *Setting up TPM and departmental committees*: TPM includes improvement, autonomous maintenance, quality maintenance, etc. When committees are set up, they should supervise and implement all those needs.

- *TPM targets:* Each area is benchmarked and the teams set targets for achievement.
- *Plan:* Create a master plan for institutionalizing; plan all the training and testing that will be done.
- *Opening ceremony*: Hold a ceremony and invite all stakeholders, including suppliers, related companies, and affiliated companies, including our customers, sister concerns, etc. Some may learn from us; others can help us. Customers will get the message that we care for quality output.
- *Implementation:* The plan is put into play. Of these, some activities will establish systems for production efficiency, for control systems of new products and equipment, for improving the efficiency of administration, for greater safety, or sanitation in the work environment. Be sure to insert structures into the management reporting and incentive system that will keep TPM in place.
- *Institutionalization:* If everything is running smoothly, you are now in the institutional stage. The reporting and incentive structures should keep the system operational. Then ask your people where to go from here. Consider striving for a prize such as the Deming Award or PM Award. Think of ideas to challenge your teams and keep everyone motivated.

Imants BVBA's Approach

Exhibit 11-1 summarizes the four phases that Imants BVBA uses to outline TPM installation. Their steps are very similar.

Exhibit 11-1 Imants approach to installing TPM

Preparation	Announcement to introduce TPM. Begin basic training in TPM. Conduct a TPM promotion tour for different stakeholders. Set up and communicate basic TPM policies and goals. Create a master plan for TPM.
Kick-off	Invite customers, affiliated companies, and subcontractors.
Implementation	Develop an equipment management program. Develop a planned maintenance program. Develop an autonomous maintenance program (see below). Increase the skills of operators and maintenance. Develop early equipment management program.
Stabilization	Perfect TPM implementation; raise TPM levels.

Seven Steps to Reach Full Autonomous Maintenance

Before the TPM process starts, educate all the employees about TPM (using the appropriate level of information for each person's role in supporting TPM). In the presentation, review its advantages, disadvantages, and steps to go through. Educate employees about abnormalities in equipment and in basic maintenance before or as this step unfolds.

The operations team has pitched management and they are on-board with the idea of TPM. The initial TPM steering committee has explained the program to all the stakeholders including maintenance, the rest of operations, stores, engineering, and others. There are seven steps to the development of an autonomous maintenance program at the individual worker level after the idea has been rolled out and universal training has begun.

1. Initial leaning, review of entire machine, tightening

This step takes place after all the dog and pony shows to all stakeholders, after top management sign-off, after a pilot area is chosen, after full TPM conceptual training, and finally after area team meetings.

In the pilot area, decide on a machine and agree on where to clean and to what depth. The first cleaning will force the TPM team to deal with the machine on a very close basis. Some use the phrase "clean to inspect" to denote this. Repair any deficiencies that become apparent during the complete cleaning. Clean and paint areas where this would be useful. Tighten all fasteners to specification. Review the entire machine operation. This is a hand-on-the-machine procedure.

- Have an initial meeting and initial review of O&M manuals, cleaning materials, tools, and goals for the activity.
- Review the PM lists from maintenance. Include a review of all frequencies of PM. The view here is to know what is happening, and not yet to change it.
- Obtain and lay out all cleaning materials and tools needed for cleaning.
- Obtain and layout all tools needed for partial disassembly.
- The group itself should assign roles. Operators should clean the equipment completely with the help of the maintenance department. Dust, scrap, rust, shavings, stains, oils, and grease must be removed.
- Remove everything from around the machine that does not belong there. As in 5S, find a place for everything.

- In addition to cleaning, this initial review should deal with oil leaks, loose or frayed wires, worn belts, missing and loose nuts and bolts, and worn out parts.
- All grease fittings should be cleaned and, based on the O&M manual (specifications on the machine), greased. Make a drawing of locations, quantities, types of grease needed.
- At some point with an expert, rationalize the usage of greases and oils to use as few as possible without compromising the effectiveness (standardize on the "better" one).
- Paint areas where this would be a benefit.
- After clean up, problems are tagged. Use one color tag for problems that can be solved by operators (if they cannot be solved right then and there). Use another color tag where support of maintenance is needed. Additional tag colors might be used for engineering, safety, or environmental input.
- Contents of tag are transferred to the CMMS or log book.
- Make note of areas which are inaccessible or difficult to clean.
- Finally, close-up and run the machine.

Exhibits 11-2 through 11-8 are handy guides that are adapted from the work of Lakshmi Narayanan.

Exhibit 11-2 is derived from the first step of the concepts of Jishu Hozen, which is Japanese for Autonomous Maintenance. It implies that the operators themselves maintain their machines in highest standards. These operators are trained to do so through the structured approach of the 7 steps of Jishu Hozen. Each step has a focus, an activity to be carried out, and certain results to be achieved.

Exhibit 11-2 Introduction to the machine -Step 1 of Jishu Hozen

Steps	Category	Activities	Target for Equipment	Target for Operator	Guidance and Promotion
1	Initial cleaning and inspection	Complete elimination of dust and dirt, especially on the equipment	Prevention of forced deterioration caused by dust or dirt	Formation of close relationship with equipment through touching and handling equipment	Pointing out and guidance of the priority clean-up areas.
		Performance of lubrication and retightening, discovery and rectification of slight equipment defects	Discovery and rectification of latent defects through clean-up	Cultivation of the ability to identify equipment problems	Instructions in the importance of clean-up (education)
		Removal of unnecessary things, orderliness and tidiness of tools and jigs (1S and 2S)	Discovery of areas where clean-up is difficult	Recognition of the importance of clean-up	Preparation of diagnosis sheets
			Removal of unnecessary things around the equipment		Responsibilities in the operation and implementation of activities
			Rationalization of lubrication		

2. Maintenance prevention and maintainability improvement

TPM is a first cousin to Lean, so every effort is made to reduce the time and materials to do the tasks. The goal is to have tasks that are effective while reducing the inputs to accomplish the task. Time is an input. Reduce time to perform cleaning. Better yet, remove the need for cleaning by removing the source of contamination or redirecting it to a trash can! This is called maintenance prevention (Exhibit 11-3).

Make the machine easier to service (lubricate, tighten, clean, adjust). Ease of service comes under the maintainability improvement. Hard to get to areas of the asset (for cleaning or lubrication) must be made easy to reach. For example, if there are many screws to open an inspection door, hinge the door and put hardware on it, or mount a (transparent) acrylic sheet.

- Hard-to-reach zerk fittings can be piped to a safe location outside the machine or automatic lubricators added.
- The machine should be modified to prevent accumulation of dirt and dust.
- Viewing ports and sampling doors should be added where appropriate to facilitate inspection and cleaning.

Exhibit 11-3 Reducing maintenance (Maintenance Prevention)

Steps	Category	Activities	Target for Equipment	Target for Operator	Guidance and promotion
2	Countermeasures for the causes of forced deterioration and improving to hard-to-access areas	Implementing countermeasures against such sources as dust and dirt, and prevention of spilling leakages.	Make cleaning and inspection easy by minimizing sources of dust and dirt and areas where inspection and cleaning are difficult	KAIZEN of nearby items to practice and master the application of the KAIZEN methods and concepts	Concept and practice of equipment Kaizen
		Implement countermeasures against areas where cleaning and inspection are difficult (shorten the time needed for cleaning and service)	Improve maintainability	Enjoy the KAIZEN activities (Pleasure of hand-made)	How to prepare the criteria or standards
		Ranking the priority places for daily inspection			Implementation of visual control and instruction on device development

3. Establish consistent standards

Specify all tasks and frequencies (daily, weekly, every 1000 pieces, etc.). Set standards for tasks (how clean, what to use to clean, how much and what type lubricant). The specifications must be direct, easy to understand, and completely reflect what the operator is doing (Exhibit 11-4). The autonomous maintenance team itself prepares documentation (with possible staff support). Wherever pictographs, photographs (simplified), drawings (iconic) can be used, they should be adopted.

- Standards for task list items should include details like when, what, and how (such as **How to** clean **what part** of the machine with **what tool** and with **what cleaning** material).
- If technique is important, it is included. Also it is important to know when you are done.
- Schedule should be made for cleaning, inspection, lubrication, and other tasks.
- Once set, the schedule has to be followed.

Exhibit 11-4 Setting Standards

Steps	Category	Activities	Target for Equipment	Target for Operator	Guidance and Promotion
3	Preparation of tentative JH standards	Tasks to allow positive clean-up, lubrication, and retightening maintenance within shortest time	Keeping up TLC clean-up, lubrication, and retightening	Self-decision of criteria and its strict observance	Preparation of standards for technologies and techniques
		Kaizen of inspection method and visual control		Each worker learns to be aware of individual roles.	Clarification of procedures to study what the equipment must be included

4. Inspection and maintenance

Initial inspection and cleaning is complete. The next step delves into all the information known about the asset and follows manufacturer's manuals, engineering recommendations, and equipment history from the CMMS (what has failed in the past)(Exhibit 11-5).

- The group is taught how to correct minor defects themselves.
- Basic training for maintenance has reached a point where the TPM teams can use their own knowledge base for problem solving.
- Review the Bill of Material for the asset and compare that to the stock list. The team, with input of parts usage data from the store room, should decide if action should be taken.
- The employees are trained in disciplines like pneumatics, electrical, hydraulics, lubricant and coolant, drives, bolts, nuts, and safety. This is necessary to improve the technical skills of employees.
- Use inspection manuals correctly.
- Employees should share knowledge with others.

The operators are now well aware of machine parts.

Exhibit 11-5 Inspection and Maintenance

Steps	Category	Activities	Target for Equipment	Target for Operator	Guidance and promotion
4	General inspection	Master inspection skill, utilizing inspection manuals	Restore deteriorated areas by exterior inspection of equipment and reliability improvements	Learn inspection skills	Prepare training texts for general inspection
		Discover and restore general inspection	Investigate areas where remedial action and normal inspection are difficult	Understand the equipment functions and mechanism	Plan education and training schedule
		Prepare autonomous inspection standards	Make inspection efficient	Master how to put together data and how to use it	Execute leader training
				Activate through KAIZEN activities	Follow up with education and training
				Learn the importance of education and communication	Prepare general inspection manual and check sheets

5. Autonomous maintenance

Inspection is turned over to the group (Exhibit 11-6). Check sheets are utilized for all inspections. Minor repairs are completed. Maintenance is only involved in major problems that involve specialized knowledge, skills, or contacts.

- Maintenance periodically reviews processes with an eye toward maintenance improvement.
- Maintenance provides audit and advice.
- TPM teams consult with experts within and outside the company to discover better methods. New methods of cleaning and lubricating are researched, adopted, and used.
- Each employee prepares his own autonomous schedule in consultation with the team and supervisor.
- Based on experience and history, parts of the asset which never present a problem or which don't need any inspection are removed from the task list permanently.
- Buy the appropriate quality of machine parts. Make sure the parts are balanced, finished correctly, and robust enough for the application. This avoids product defects, shortened asset life cycles, and potential breakdowns.
- The frequency of cleanup and inspection is reduced, based on experience. However, daily inspection points should not be neglected.

Exhibit 11-6Autonomous inspection is an important part of Autonomous Maintenance

Steps	Category	Activities	Target for Equipment	Target for Operator	Guidance and promotion
5	Autonomous inspection	Ongoing review cleaning, bolting, lubrication, and general inspection criteria; constantly work to increase efficiency	Positively keep restorations from deteriorating by means of general inspection		Teach how to precisely analyze the data
		Prepare and implement autonomous inspection check sheets	Analyze equipment that has good operability for additional improvements	Maintain one's own equipment by oneself	
		Improve visual control and operability		Self-decision and strict observance	Effective equipment management and maintenance
				JH is learning what's the ideal	

6. Organize to support ongoing TPM efforts

Enthusiasm, special improvement projects, autonomous teams will only go so far. For TPM to become real and stay active it has to be supported by the systems that run your business. These systems remind people to keep focus on improving OEE which in turn keeps people engaged.

- Use structures (reporting, incentives, awards, recognition) in the company to support TPM.
- Systemize the autonomous maintenance activity.
- Align the organization to support TPM.
- Use TPM productivity reports to run the plant.
- Continue to develop and publish standards for all activities.
- The environment the machinery sits within should be organized.
- Necessary tools and supplies should be organized so there is no searching or search time is reduced.
- Work environment is modified such that there is no difficulty in getting needed items.
- Everybody should follow the work instructions strictly.
- Necessary spares for equipment have been planned for and procured.
- OEE, OPE, and other TPM targets must be achieved by continuous improvement.

Exhibit 11-7 Organize business systems to support TPM

Steps	Category	Activities	Target for Equipment	Target for Operator	Guidance and promotion
6	Standardization of systems	Pick items to be managed	Review equipment reliability, maintainability, and operability	Improve management technology	Technical guidance to promote standardization
		Standardize management items; systematize maintenance managements	Review equipment, environment, and layout	Expand JH management scope	Revision of management standards and thoroughness of management
				Thoroughness of visual control	

7. Full functioning TPM

Once all the steps are accomplished there is now a program that will sustain itself (for a while at least). Continuous improvement means just that, it never stops. You will have to fight to keep people engaged and not revert to old habits. To keep people's interest up some firms apply for the TPM prize.

- Track the results of the effort and give ongoing recognition to progress.
- Monitor failure frequency and look for additional improvements.
- Spend more time on improvements that reduce maintenance effort while increasing equipment availability.
- Consider applying for a TPM prize.

Exhibit 11-8   Full Functioning TPM

Steps	Category	Activities	Target Equipment	Target Operator	Guidance and promotion
7	Autonomous Management of TPM	Develop company policy/goals and quantitative analysis activities	Improve equipment reliability, maintainability, and operability by applying various data analyses	Enhance goal consciousness, including maintenance cost	Technical assistance for equipment
		Positive implementation of MTBF analysis and recording (recording failure by visual control)	Keep overall equipment efficiency at its best	Acquire increasing skills to perform minor repair by oneself	Standardization of improved items
				Access data recording, and analyze techniques, and use best technology	Education and training in repair skills

What is needed?

Enrique Mora, President, LeanExpertise.com (35 years of experience includes supervision of manufacturing and design processes in Mexico and the U.S., and implementation of lean manufacturing, Six Sigma, and TPM http://www.tpmonline.com), reported the factors that promoted success in TPM installations:

- Strong support from top management
- Leadership in the implementation teams
- Follow up on the Continuous Improvement (Kaizen) processes
- Full time coordinator
- Full time PM technicians
- Education and training for operators so they can do Autonomous Maintenance
- Development of an implementation plan and then following it
- Steering committee complies and preserves their goals
- Continuous motivation
- Displays show details and results of the projects
- Distinctive recognition of participants
- Rewards program in place

Key Concepts for the TPM Installation

Training

Organize training, training, and more training. TPM knowledge has to be distributed throughout the organization. There are several levels of training necessary. Larger organizations will designate TPM warriors, Samurais, or black belts who have had the most in-depth training from a variety of outside sources and readings. They will conduct a feasibility study as well as conduct basic training. Smaller organizations might use a consultant for this role.

Measurements

The measurement benchmarking phase establishes the baselines (from your existing production numbers and from associated budgets) against which all future improvements will be measured. It is essential to pinpoint current operation ratios so that later you can prove that TPM has helped you.

Pilot

With many new programs, the smartest way to train people and get them interested in the program is through a trial run or pilot program. Choose a small area to release the program, work out the kinks, and create your first successes.

The pilot area should be small enough to have a good handle on all activities and large enough to make an impact. It should be typical of the other areas. Pick an area that has well-known physical boundaries (to reduce confusion) and where the people are interested and excited about change. The people in the pilot area will be dispersed to other areas once the project is proven.

The problem here is to stay on top of the pilot so that any problems that come up will be handled, rather than crashing the program. Because you have set up baselines in the benchmarking phase, you can use these early results to demonstrate the potential.

Roll out

Roll out to more, most, or all of the plant, depending on the results thus far. A phased installation is usually advisable; this way, key team members can start their training on operational teams in addition to formal classroom training. This rapidly expands training capability and effectiveness.

Structures to keep it going (Sustainability)

Generally the problem is not making some change. The problem is making the change stick after top management drifts off to slay other dragons. The process of making the TPM transformation stick is called sustainability. It is critically important.

There are various ways to keep the installation on track. However, the most important way is to report on TPM indicators in every production and maintenance meeting. Once TPM achievements have become part of the existing reporting structures, the plan has an excellent chance of longer-term success.

Outside audits and visits

A great idea is to use outside people to conduct a more in-depth look at the TPM installation on a periodic but not frequent basis. Large organizations might invite TPM Samurais from other plants to be the outside "auditors." Otherwise, consultants serve well in this role. The mission is more than pointing out what's wrong; it can be used to point out any opportunities missed.

Go for the gold

You might choose to apply for a TPM certification from the International TPM institute. This may be especially important to your customers in certain industries. By all means, throw a TPM party. Acknowledge the success stories. Perhaps the team can roast the leadership and have some fun.

Look Out: Problems Ahead! Why Do Half of all TPM Installations Fail?

If you think about the tall order being asked for, it is a surprise that one out of two installations succeeds! The failures hurt the ability of the organization to make any big change. The workers get the idea that when the going gets tough, their leaders run for cover.

The reasons vary by company.

- Lack of understanding of TPM
- Thinking that TPM is a maintenance program
- Focusing on the maintenance aspect of TPM and not the awakening provided by TPM
- Lack of buy-in by Operations
- Lack of ownership by operations management
- Lack of operational leadership
- No structure (sustainability) to support ongoing program
- Old incentives still in place
- Lack of commitment for a long-term project
- Change in leadership with new priorities
- Not enough resources
- Unrealistic expectations

Enrique Mora (mentioned earlier) reported in his experience the factors that prevented success:

- Only the Maintenance group got involved
- Implementation teams never met again
- No further improvements took place
- The people in charge of TPM has other functions that took most of their time

- Maintenance technicians devoted 50+% of their time to Emergency Corrective Maintenance (firefighting mode)
- Operators never received enough training on Autonomous Maintenance
- No master plan was created or, if it was, it was not complied with
- No formal steering committee meetings were regularly held
- Minimal or no motivation

Any implementation of TPM will face real problems. Following the plans already mentioned will minimize the negative effects. At a TPM seminar recently, supervisors were asked what real problems they imagined they would encounter installing TPM. Some of their responses are below. These problems must be considered, discussed, and overcome to have an effective TPM effort.

- Top management sign-off and support throughout a multi-year TPM process. If your management has a short view, they might agree to a multi-year plan and withdraw their support after the first year.
- Top management might give lip service but does not support it with their deeper commitment and their time. How do we get top management to be boosters of the program?
- Many people treat it as just another "Program of the month" without paying any focus; they also doubt its effectiveness.
- TPM is not a "quick fix" approach; it involves cultural change to the ways we do things.
- Insufficient resources (people, money, training, time, etc.) and assistance are provided.
- Supervisors might criticize rather than solve problems; they

also might complain or undermine the success of the project in front of their subordinates.

- Middle management doesn't really understand TPM, either practically or philosophically.
- Some people at all levels may oppose the change (it was good enough for my father so it's good enough for me) or lack under standing of the need for change.
- TPM requires some minimal downtime. How do you integrate TPM with customer demands and the sometimes unreal demands of the forecast and production schedule?
- The workers might object to the perceived "extra" work with a slow down, increases in absenteeism, letting quality suffer, etc.
- Workers might fear job loss; as they become multi-skilled, there may be a need for fewer people.
- Where does an organization — whether small, medium, or large-sized — get time to do all the training necessary?
- Departmental barriers exist between maintenance and operations, limiting cooperation.
- How do you run TPM in a high-turnover situation? Operators don't stick around long enough to get trained.
- Where do temps fit into TPM? We use temps to operate machines during busy times.
- Who really manages the maintenance part of the operator's job? How do we get Operations to take advice about operator maintenance effort when they won't now?
- Is there willingness and interest to accept the new roles of the two groups?

- Operators don't like to clean equipment and neither do maintenance people.
- Where do our die setters (set-up people) fit into this scheme?

Chapter 12
Training

For TPM to Work, You Better Be Great at Training!

On one level, TPM is really about new skill sets and the application of these skill sets. TPM requires significant additional training for the operators. They have to be brought from their existing knowledge level to a level that will assure that they can understand and successfully complete the TPM tasks. Many TPM efforts fail because they do not allow for sufficient training.

Think about asking someone without a lifetime of maintenance experience to do basic maintenance tasks and solve problems. Some of the background knowledge is not there and cannot be assumed.

Exhibit 12-1 examines your current overall training organization on the following page.

Exhibit 12-1 Training Practices

Quick questions for your current training practices Is there a present system of education and training for operators? Do you have a written SOP with steps for educating and training activities? Are there set policies, priorities, budgets, and goals? Are employees trained for upgrading the TPM skills determined by analysis? Is training based on task analysis and existing candidate competencies? Are candidates routinely post-tested to be sure materials were learned? If not, adjust approach. Review existing certificated operator programs or operator levels that exist. Is there an ongoing evaluation of activities and study future approaches?	
Actions to take Prepare training calendar if one is not present. Establish training system for operation and maintenance skills if one is not present or not adequate.	

Three Competencies Create Three Training Issues

This section is adapted from the author's 2009 book Handbook of Maintenance Management, 2nd edition.

In training there are three domains of learning: knowledge, skill, and attitude. Many types of training address one of these types of learning without regard for the other. Maximum effectiveness must come from competence in all three areas. You can see this particularly clearly in TPM.

The operator needs specific skills, some knowledge, and probably most importantly the 'right' attitude. An operator with the skills and some knowledge but without the 'right' attitude is a problem. Too much confidence and the TPM operator will undertake tasks inappropriate for their competence level (and possibly put themselves and others at risk). Too little confidence and they will only undertake TPM with reluctance. In either case, effectiveness is compromised.

KNOWLEDGE

Type	Observable behavior	Performance level
1. Knowledge	Be able to describe diagram, argue, etc.	Answer X of 10 questions correctly

Examples of this domain include questions such as: What is the process to heat treat steel to a particular specification? Can you describe the steps to obtain a hot permit in this facility? How would you program a cascade pumping arrangement with an Allen Bradley PLC?

Knowledge is what is taught in most schools and is a building block of education. Generally it is easy to test in this domain. Sometimes you will find mechanics and TPM operators who can do some of the work, but don't know what or why they are doing it. When people know how to do something without knowing what they are doing, two problems are created. The person might unintentionally create a dangerous situation or a cascade failure in another system. The other problem is that, without the knowledge, the person cannot be as creative and think outside the normal way to solve problems.

SKILLS

2. Skills	Demonstrate, show, perform, solve	Do ... in x minutes with no mistakes

Example of this domain include asking candidates to demonstrate their competence by welding two pieces of 308 stainless in a vertical position, or to make a MI cable connection to at least 3 mega ohm resistance to the casing. Most on-the-job training consists of skill training. In maintenance, we admire skilled mechanics. If you have the proper demonstration set up, competence is also easy to test. One trap in TPM is that we skip the knowledge on the presumption that operators don't need it. Operators need enough of the knowledge to see why they are doing what they are doing and how it impacts the whole process.

Many people have the knowledge without the skill. We say, they can talk a good game... but can they weld, etc? Metallurgists might be people who can explain the details of the welding process without actually knowing how to weld. They could specify the tools, heats, covering gases, and wire type, without being able to do the welding themselves.

ATTITUDE

3. Attitude	Comfort, without hesitation	To your own satisfaction

Examples of this domain include discussing the operators' comfort level with a particular technology or their work ethic. This is more difficult to test. Some could conceal their discomfort. You occasionally run into trades people who have the skills and knowledge, but lack the will or the confidence to do the work.

Attitude problems might start with a void in a skill or knowledge that creates frustration and a bad attitude. In other cases, the person might lack capabilities such as strength or height to do the job. Fear might also play a role.

A fourth attribute cannot be addressed by training or education. Some people might not have the aptitude to be TPM operators. TPM requires a certain amount of intelligence, certain levels of strength, flexibility, and visual acuity, and it helps also to have a desire to advance. It is important to weed out people who do not have the minimal aptitude.

Education Versus Training

There is a difference between education and training that applies particularly to TPM. As defined in Wikipedia (both definitions), training is generally associated with specific skills and techniques. You can be trained to weld or how to wire instrument panels. Education concerns the whole person.

In TPM, we seek to educate the operators as well as train them. The reason is that we want TPM operators to be able to think through the problems and not just apply the predigested solutions we give them.

Education in its broadest sense is any act or experience that has a formative effect on the mind, character, or physical ability of an individual. In its technical sense, education is the process by which society deliberately transmits its accumulated knowledge, skills, and values from one generation to another through institutions.

Teachers in such institutions direct the education of students and might draw on many subjects, including reading, writing, mathematics, science, and history. This process is sometimes called schooling when referring to the education of youth.

The term training refers to the acquisition of knowledge, skills, and competencies as a result of the teaching of vocational or practical skills and knowledge that relate to specific useful competencies. It forms the core of apprenticeships.

One can generally categorize such training as on-the-job or off-the-job:

- *On-the-job training* takes place in a normal working situation, using the actual tools, equipment, documents, or materials that trainees will use when fully trained.
- *Off-the-job training* takes place away from normal work situations. Off-the-job training has the advantage of allowing people to get away from work and concentrate more thoroughly on the training itself.

You go to a university to get an education and you go to a trade school to get training. Of course the two categories overlap. The effect of this distinction on TPM is clear. Some topics, such as reliability, are better dealt with in an education framework and others, such as tightening bolts, in a training framework. They also have implications for the act of training. Training is done either on the job or off the job whereas education is generally off the job.

In the lists below, many topics are partially shared. For example, safety is an area of education whereas safe work practices are considered training. The former involves theories, attitudes, and background to help understand the whys; the former is concerned with detailed practices.

Education Topics Provide a Background or Context for TPM

- RCA (Root Cause Analysis)
- Reliability
- Quality
- Safety
- TPM (background, meaning and context)
- TPS (Toyota Production System)
- Life cycles of equipment and mechanisms of breakdown

- How your production works, how your industry works
- Preventive Maintenance
- Maintenance Management

Training Topics Provide Specific Skills and Knowledge to Run TPM

- Tightening bolts (see appendix for one such training program)
- Lubrication (see appendix for one such training program)
- Cleaning
- Inspection (detailed look at different failure modes)
- Measuring
- Sketching
- CMMS usage and filling out corrective work orders
- Specific adjustments
- Safe work practices
- Tool use
- Inspection of product
- Team participation
- Minor repairs

A simple and useful guide to the different levels of mastery of a skill or knowledge area:

Phase 1: Do not know.
Phase 2: Know the theory but cannot do.
Phase 3: Can do but cannot teach
Phase 4: Can do and also teach.

Effective Training

In Japan, operators will go through as much as 12 weeks of training before they are let loose to join as junior members of an autonomous maintenance team. In addition, they would expect significant on-the-job support for the first few months.

For TPM to be successful, significant training investments must be made.

The most successful training technique is a structured 4-part flow. As you analyze the individual jobs you will be building up a library that can be reused.

Step 1: Job Analysis

Look at the job to be done. Determine what knowledge, skills, and attitudes are needed for the job. Before we can look into teaching anything, we have to see what is needed. Look at the job as it is today; forecast where the job is going in the short term. The big picture for competencies is called the General Learning Objective (GLO).

The concrete and specific skills, knowledge, and attitudes required to do the job are called specific learning objectives or SLOs. If properly designed, a person achieving these SLOs will be successful in this job.

Step 2: Candidate Evaluation

Evaluate the trainee's (or the trainee group's) current skills, knowledge, and attitudes; then compare them to the results of the job analysis. This evaluation is targeted at a specific set of competencies (such as lubrication). It is not like an annual evaluation.

A direct supervisor might be able to make an educated guess. If the trainees have good insight, they might know where they are weak. Most evaluation situations require some kind of testing (either observation on the job or more formal written or bench tests). Testing should be designed to uncover the skills, knowledge, and attitudes on your job analysis list. Testing should also un¬cover those people who do not have the aptitude to learn the new information.

It is important to note that success on the test should correspond to success on the job. Testing that does not reflect job requirements is said to be invalid. The Americans with Disabilities Act (ADA) and related legislation are clear that the test must not discriminate against any group, disability, or condition. The test should also reflect actual job conditions. For example, if the worker must lift 100 pounds in the test, the actual job must call for heavy lifts where equipment cannot easily be used. ADA banned arbitrary tests that were not closely related to job success.

Step 3: Fill the Voids

Translate the voids in skills, knowledge, and attitudes of the potential trainees uncovered above in order to develop a training lesson plan. The training plan should list all of the types of learning that they need.

Training prescription summarizes the skills, attitudes, and knowledge that the candidates lack which are needed for the job. The estimated time requirements, costs, time off for the trainees, and any requirement for supporting staff would be developed from the training prescription.

Step 4: Post Test

Retest the trainees and determine that the voids in skills, knowledge, and attitudes have been filled. Many training efforts complete the first three steps and never validate if the learning objectives have been met. Increased rigor in this area will make it easier to choose effective training and prove the case for additional training resources.

Certified Operator and Certified Mechanic Training

It is impossible to assume quality will result when the people interacting with the asset don't know their job. Donald Overfelt, formerly a supervisor for Lucent Technologies, explained his company's methodology for a certified mechanic.

They took each machine and broke it up into components. For each component they thought of all of the action words (such as clean, adjust, lubricate, replace, rebuild, align) and created a matrix first for each machine, then for each mechanic (Exhibit 12-2).

Exhibit 12-2 Machine Matrix: Activities that can be done on this asset

Component	Clean	Adjust	Calibrate	Rebuild	Replace
Proximity assembly	X	X		X	X
Limit switch	X	X			X
Main shaft and bearings	X	X			X
Component x, y, z	X	X	X	X	X

Then for each activity the question is asked, "Who has the competence?" Everyone who interacts with the machine is tested or evaluated. Each operator on the TPM team and each mechanic would be required to be trained and tested in each area of competence.

Exhibit 12-3 Mechanic matrix: What activities is this individual certified on?

Component	Clean	Adjust	Calibrate	Rebuild	Replace
Proximity assembly	C	C		X	C
Limit switch	C	C			C
Main shaft and bearings	C	T			T
Component x, y, z	C	C	X	X	C

C—Certified	X—Not certified yet	T—In training

An operator matrix for the same asset would focus on "clean and adjust" boxes and stay away from calibrate, rebuild, or replace (unless the asset was particularly straight forward)(See Exhibit 12-3).

Categories for Sources of Training

1. YOUR STAFF IS YOUR FIRST CHOICE OF POTENTIAL TRAINERS.

Within your staff there are several possible opportunities for train-ers. Please note that being a trainer should be viewed as a job-enhanc-ing project. Time should be given for preparing materials. The trainer should be relieved of other duties.

- Tap your soon-to-retire workers as trainers. This group has significant experience that should be channeled into the next generation. In some organizations, people who have already retired are recruited to return as part-time teachers.
- Use the internal guest instructor concept. With this concept, a staff member would be treated as a guest (receiving lunch, clerical support, go off-site for longer trainings, get relief from other duties).
- Tour training is an excellent team-building exercise. Once a month, you tour a section of your facility and the most experienced person plays "show and tell" about the problems and successes in his/her area.
- Video technology has rocketed ahead so that most firms can afford a quality video camera and editing software. New equip ment set-up, construction documentation, and specific machine training are popular first subjects. Craft training is a more difficult but rewarding area. After expertise is obtained, any topic can be recorded to good effect; these can then be used to train new operators and mechanics.
- Look to other parts of the organization such as human resources, data processing, engineering, and production for expertise useful to your upgrading effort.

2. TRAINING IS BIG BUSINESS FOR A LARGE NUMBER OF ORGANIZATIONS.

Here are some ideas:

There are many excellent companies that provide TPM or craft training. These firms can provide professional instructors, testing rigs, and video (either streaming or DVD) or audio (also available in different formats including streaming over the Internet, MP3 for IPods, and traditional discs). The quality and appropriateness to your operation may vary, so check several vendors.

- In-house courses are available on a wide variety of maintenance topics. Most appropriate if many people need the same training. Costs are about $1500–$4500 per day.
- Public seminars are useful for training 1–3 people. Expect seminars to cost $500–$3000 and last 1–5 days. Try to get recommendations of the better seminars from people in your industry.
- Many organizations sell DVDs. Expect to spend $50–$7500 for video training in a wide variety of areas.
- Interactive computer or web-based videos can lead a trainee through a series of lessons and retrain when necessary. Expect to pay $250–$500 for a rental and up to $8000 to purchase a course. These typically require only a standard PC (though the experience will be enhanced with an extra nice monitor and speakers).

3. EQUIPMENT MANUFACTURERS.

This is a growing area. Equipment manufacturers have a vested interest in a trained user base. Many of them subsidize training by calling it a marketing expense. Time and time again it has been shown that trained users are happier users. Try to negotiate training into all equipment purchase contracts. Excellent low-cost training is usually available from vendors of predictive maintenance hardware.

4. TRADE AND PROFESSIONAL ASSOCIATIONS.

These groups strive to increase their value to their membership. One of the traditional ways is to provide industry-specific training in either traveling seminars or at workshops during trade shows. If your association does not provide training that you believe is needed in your industry, why don't you volunteer to put a seminar package together for the association?

5. INTERNET TRAINING.

Some computer networks provide internet-access training sold by the hour. These systems train in electricity/electronics, pneumatics, building trades, business subjects, computer subjects, basic science, and many other areas.

6. TECHNICAL, VOCATIONAL, AND CAREER SCHOOLS

Tech schools are an excellent source for trade training. Get to know the people running your local tech schools. Visit and walk through the facility. Many companies set up specific labs, benches, or workshops with equipment that the tech school uses to train students. Many tech schools are willing to negotiate training contracts for some or all of your technical training needs.

7.   COMMUNITY COLLEGES, COLLEGES, AND UNIVERSITIES.

These educational institutions are frequently looking for new markets. They have significant expertise in teaching more advanced subjects to adults. Many of them have entered into instruction contracts with private industry in areas including computerization, robotics, regulation, automation, and business skills.

8. UNIONS

Some unions are rethinking their traditional roles. Many see that skill needs are shifting and have decided to lead the trend by setting up training for their members. This might be an interesting subject to raise if your union is not doing this already.

9. INSURANCE COMPANIES.

Insurance companies can cut claims by conducting certain types of training. Some firms will send risk managers through your facility and provide specific training in areas such as safety, risk management, liability reduction, fire safety, storage and handling of chemicals, record keeping for maintenance, safety, and accidents.

10. GOVERNMENT.

Government agencies provide seminars and workshops on a wide variety of topics including EPA issues, hazardous materials, waste disposal, safety, record keeping, and dealing with overseas vendors.

Methods to Consider for Training

- Coaching, OJT (On the Job Training): 1-on-1 training and encouragement (especially good if you use the technique TSED — Tell, Show, have them Explain, have them Do). This technique is suitable for teaching important new skills where the high time investment is justifiable.

- Case method: Analyze a specific incident, problem, situation, or company.
- Correspondence: Home study of commercially-produced lessons. They can be adapted to training within a firm. In many cases, the materials can be used again.
- Books and reading materials: Low-cost method, good for people already skilled to add a specific skill or knowledge area. Permanent, can be referred to in the future and used by others.
- Video and audio discs: These have the advantages of lecture, books, and demonstrations combined. Permanent, can be referred to in the future and used by others.
- Conference: Send someone to a public conference with a training program. Trainees can get exposure to many instructors, peers, and vendors at the same time.
- Cross-training: Training someone skilled in one area with another skill or craft.
- Demonstration: The best trainers tell the trainee what they will learn (such as welding), show the trainees how it is done (by welding in front of them), and finally tell them what they saw. Some skills can be taught just by demonstration, particularly when the trainee is already skilled in related areas.
- Laboratory: Experiments designed to teach by discovery.
- Lecture: Trainers directly instruct trainees with the material to be learned. Provide basic information on a topic. Lectures can be used to introduce topics to many people at the same time (such as what is TPM).
- Programmed learning: Trainees go through material at their own speed. This training can be accomplished through books,

disks, or on-line lessons. Accommodation can be made for trainees who need additional material in some texts.

- Role play: Trainees play a role in a simulation of real situations, and learn by doing as well as through the reactions of the other role players.
- Simulation: Trainees are presented with a realistic scenario and then work alone through problems and situations.
- Distance training: Trainees use a computer and Internet hookup to conduct training from a remote location. Several plants or buildings can share a great instructor at the same time.
- Distance degree programs: These programs combined several different channels for learning and are managed from a portal for the program and class.

Considerations for Running the Training Itself

Logistics for training in the modern organization can be complicated. There are hundreds of details to manage (failure in any one of which could ruin the event). To manage any activity there are three things that must be known. First is the activity itself, second is who is responsible SPA (Single Person Accountable) and finally the "by when?"

It is important that someone is responsible (SPA) for all activities and we need to know when it is OK to bug them (before or after the next action date) to see if the activity is complete.

Exhibit 12-4 Check List for In-House Training

Consideration (be specific)	SPA*	Next action date
Materials		
Staff (inside, outside room)	✓	March 12
Outside firms needed		
Trainees		
Structure (agenda)	✓	March 15
Training Aids		
Facilities		
Accommodations		
Food and refreshments	✓	April 9
Dates		
Travel		
Promotion	✓	March 20
Timing		
Add others…		

CONSIDERATIONS FOR EACH ITEM ON THE CHECK LIST IN EXHIBIT 12-4

- Materials which includes books, work books, hand outs, and worksheets.
- Scheduling and coordination of staff time, including contingency plans if key players cannot attend.
- Check dates for conflicts with vacations, holidays, local holidays, hunting seasons, work schedules, and bad weather.
- Replacements needed for staff in training on the shop floor. Consider people needed both inside and outside the training

room, before and after the training.

- Outside firms needed for guest speakers, professional trainers, turn-key training, slides, video production, audio taping. This includes purchase orders, coordination, detailed directions and exchange of information about room set-up.
- Trainees were invited and have sent in their RSVP. Someone might have to bug them or their supervisors about attendance. Be sure to avoid shutdown days where everyone is needed, or the days after a shutdown where everyone is catching up on sleep.
- Have a plan for responding to various levels of emergencies so that everyone doesn't get disturbed for every little breakdown. It would be great if all the cell phone calls can be forwarded to a central communications center.
- Structure of training might include break-out sessions (and room to move around), hands-on bench work, and access to computers and classroom.
- Aids including Power Point, flip charts, microphones, slides, videos, projectors, screens, satellite link-ups. For the best learning conditions, include comfortable chairs and tables for writing on, perhaps even access to power outlets to plug in lap tops for taking notes.
- Facilities should be comfortable and large enough for the number of trainees expected. Person accountable for facilities should know: Who is the contact person? Who has the keys? How do you turn everything on and off?
- Accommodations are needed for trainees who are traveling in from another facility and for trainers who are from out of town.
- Food and refreshments help make people more receptive.

Remember, people have different tastes. Only 40% of people drink coffee in the morning; others may drink tea, water, juice, etc. Plan appropriate refreshments for your group.

- If people are traveling by air or train, coordinate pick-ups, tickets, vans, etc.
- Use promotional techniques to sell the program, persuading people to want to attend.
- Timing can make or break a program. A giant reorganization before a training can be a good thing, or may insure that no one will have their minds on the material. A major reorganization after the program can be devastating.

Chapter 13
Structures to Keep TPM Going (Sustainability)

The single biggest obstacle to keeping TPM going comes when the attention of management strays to some other new program or wherever there is pressure to make short term increases in output. The tendency in these cases is to return to the old ways because they worked before. They are still built into the reporting structure and are part of the culture.

The first step is to look deeply into the incentive systems (both financial and cultural) already in place before TPM. These must be dismantled. The cultural change will take time and reinforcement, partially by the new incentive structures put into place.

Also, in general how do you nail down and sustain new behaviors so that they persist?

- Be sure the routine operations reporting includes the new TPM measures.
- Give out rewards for TPM KPI (Key Performance Indicators) such as OEE.
- Develop rewards for achieving TPM milestones with money, recognition, or trinkets (such as jackets, caps, mugs, etc.)
- Remove rewards for old KPIs that conflict with TPM.
- Get plant managers involved and have TPM teams report to

them about TPM activity.

- TPM could be driven from the MRP system as a job or activity.
- Use the PM system in some way to highlight TPM compliance.

Keeping TPM on Track

Jack Roberts, a professor with the Department of Industrial Engineering and Technology at Texas A&M says in a recent article, "Without frequent assessments of plant and equipment readiness, it is difficult to achieve continuous improvement in maintenance operations over long periods of time. Unfortunately, not enough attention has been paid to this aspect of TPM."

Site inspections (or audits) are a useful tool to insure TPM is still active.

The general status of cleanup efforts in the work area is obvious to even the most casual observer. Either a facility is clean or it is not. Equipment is clean and working properly, or it is not.

Changing the physical layout of a facility in order to improve workflow may take the longest of any of the items listed. This may involve moving large equipment, purchasing new or additional material-handling equipment, or modifying the facility through re-modeling or new construction. A renovation or expansion master plan is usually created to address this need.

Developing autonomous maintenance activities that involve the machine operators in daily machine maintenance is often one of the most difficult TPM precepts to implement. Operators have traditionally been told to just operate the equipment, then if it breaks down or needs adjusting, notify the maintenance department. Under TPM, the concept of autonomous maintenance is applied, allowing operators to take an active role in the maintenance and adjustment of the equipment they operate.

Developing a useful observation and reporting system — one that not only gathers production data, but also establishes limits on certain parameters so that problems can be addressed before they become critical — usually involves all production personnel as well as maintenance and facility managers in the design process. This is where close attention to the means and methods of data collection becomes one of the major attributes associated with successful TPM implementation.

Potential Failure of the Checklist System

The development and use of checklists is certainly a necessary data-gathering activity when implementing TPM. But how do we know if the checklists are being used properly, or if they are used at all? Human nature being what it is, repetitive tasks are seen as monotonous and unimportant to many. Workers ask why they have to do the same thing over and over, the same way, day after day. Workers may get complacent or they may hurry and dismiss routine checklists as unimportant or unnecessary.

Of course, historical data cannot be relied upon if it is not gathered in a timely and repetitive manner with all of the variables under tight control. Often the only way that data gathering activities are policed is by overt observation or spot checks. Either method may be construed by the work force as yet another case of management mistrusting labor.

In Complete Guide to Preventive and Predictive Maintenance (by Joel Levitt published by Industrial Press), there is a list on how to insure PM tasks are done as designed. It is clear that some of these techniques can be directly translated to a TPM environment.

---

1. Inspect the PM work on a random basis.
2. Does the TPM inspector know how the TPM task fits in to the overall scheme?
3. Drag your top management down to the bowels of the facility and have them address the operators about the criticality of TPM and its impact on output and safety.
4. Present the job as important.
5. Have a display of accomplishment in a public area.
6. When uptime is good, make it a practice to send out letters of commendation
7. One of the most important things you can do to insure the work is done is to let your TPM operators design elements of the system and tasks themselves.
8. One hole is lack of specific skills. Be explicitly sure the operators are fully trained.
9. Implement their ideas for improvement and make the improvements.
10. If training and testing are involved, prepare and give out certificates of competence.
11. Improve the relationship between the mechanics and the TPM operators.
12. Make it easy to do tasks by reengineering.
13. Simplify paperwork.
14. Improve accountability by mounting a sign-in sheet inside the door to the equipment.
15. Make TPM into some kind of game.
16. TPM professionals like new, better toys. (Sorry — better tools, not toys!)
17. In any repetitive job, boredom sets in. Consider job rotation, reassignment, project work, and office work like planning, design, and analysis to improve morale.
18. Be sure that TPM is part of the normal reporting up the ladder in the company. Train your managers to ask questions when the numbers change.

If checklists are used, managers should consider the computer-based checklists that facilitate interaction with their front-line personnel. By accessing the experience of people who have direct, physical contact with the plant, managers can get the kind of data they need for valid and reliable assessments of equipment effectiveness.

Web-based checklists can help to gather input for TPM implementation, while also providing data for both production management and risk assessment.

Organizations committed to TPM will give serious thought to long-term condition monitoring. Equipment and systems need to be assessed frequently and correctly.

Chapter 14
Is TPM for You?

Before we discuss the topic "Is TPM for you," please understand that there are elements of TPM that apply to every type of maintenance.

How to Decide that TPM Is for You

Before developing and implementing a TPM system, it may be a good idea for the organization to research what TPM is and what it isn't. A book of this type is an excellent introduction because it is designed to touch upon all of the basic TPM issues.

Choose a group to look into TPM. This group should include leaders from operations, maintenance, engineering, and other groups. It should be lead by an operations heavyweight. Initiate a serious study of the TPM field including attending conferences, reviewing literature, and videos. Try to identify organizations that practice TPM and visit them. Build a library of TPM materials.

Conduct a quick audit of your own operation or hire a consultant to conduct the audit. Your goal is to uncover what capabilities already exist (e.g., an effective training department) and what is missing (e.g., no certified operator program). Also look at the possibility of long-term support for the program.

Present your conclusions to top management. First you'll have to teach them about TPM. (Consult Ed Hartman's excellent article in Maintenance Technology on-line at www.mt-online.com called Prescription for Total TPM Success for suggestions regarding a TPM proposal to top management.) Top management go-ahead will probably be necessary for the feasibility study and for the increased levels of training.

Is TPM feasible in your organization? Although you should be pretty sure at this point, a more concise study may be in order. This study will also diagram policies and basic procedures, and make sure there is organizational (and logistical) support for the team's initiatives. The other function is to prepare the ground for the new world-view. An outline for feasibility study follows.

TPM Feasibility Study

A feasibility study is an important element in a successful TPM installation. This agenda outlines the feasibility study process developed and used by International TPM Institute. It is adapted from Ed Hartman article from www.mt-online.com Maintenance Technology Magazine, 9/18/01.

Overall Agenda for TPM "Live-Fire" Feasibility Study

This is not your father's ivory tower feasibility study. It involves getting your hands dirty in the nitty gritty TPM innards. Once through this study, you will have a good idea of the results of the program and the difficulty of getting there.

Getting started

- Learn and practice overall equipment effectiveness (OEE) observations and calculations.
- Designate teams for the feasibility study (there may be more than one).
- Select equipment area or department to study (for feasibility study).
- Plan and schedule practice run.
- Practice OEE observations, calculations, and review.
- Enter reasons for breakdowns and idling/minor stoppages on OEE form.
- Make Pareto charts for breakdowns.
- Organize for daily input and calculations (spreadsheet) during feasibility study.
- Develop plan to summarize OEE data for all equipment in the pilot area.
- Develop tasks, schedule, and staffing plan for the feasibility study.
- Develop feasibility study schedule (typically 8 weeks).
- Establish and document the baseline (all current data).
- Collect maintenance costs (especially breakdown repair costs) and other data for baseline.

Personnel

- Carry out "skills required vs. skills available" analysis (during feasibility study).
- Develop chart of skills and check correlation to current pay grades.
- Summarize competences and develop required overall training plan.

- Determine trainability of operators; check past participation in courses and results.
- Determine motivation; interview operators and supervisors as needed.

Machines

- Develop customized forms for selected machines. Each machine has customary losses and the form should reflect those losses.
- Use failure information sheet (FISH) as a test, especially if there is no usable breakdown data. Then expand when ready to respond to operator's suggestions.
- Practice equipment condition analysis and review.
- Assess condition of tools, dies, and fixtures along with major equipment.
- Document equipment problems (dirt, rust, spills, leaks, low oil levels, loose and missing parts, etc.) with color photos.

Maintenance

- Analyze current maintenance operations.
- Assess current preventive maintenance (PM) program and results (including status of checklists, PM work orders, PM schedule, PM compliance, PM reports, equipment history, and predictive maintenance).
- Assess maintenance management in general: work order system, maintenance information system (CMMS), planning and scheduling, maintenance control (reports), organization,etc.
- Assess and report on housekeeping, cleanliness, discipline, procedures, etc.

Culture

Report on the corporate and plant culture as well as the existing teamwork. Items to be covered include:

- Level of employee involvement, enthusiasm, team spirit, etc. (Include management style, empowerment, delegation, etc.)
- Number and location of existing teams
- Function (purpose) of existing teams
- Are they still active?
- How will future TPM teams fit into the current existing team structure?
- Can existing teams (manufacturing) be converted into TPM teams?
- Can TPM be added to existing teams?

Plan the Installation

- Develop a custom TPM pilot installation plan that will include two areas (departments) at a minimum.
- Propose a number of TPM teams.
- Provide a training plan .
- Provide an installation schedule.

Proposal

The TPM coordinator develops a draft (outline) and then writes the feasibility study report (input from all teams). Include typical examples of losses and improvement opportunities (benefits).

Prepare for the feasibility study management presentation that should:

- Present all relevant data from the study in organized form.
- Provide for across-the-board participation (including operators

and maintenance).

- Propose the TPM organization (at least for the pilot installation).
- Present a draft TPM vision, mission statement, policy, and strategy.
- Propose a TPM logo (and motto, if appropriate).
- Present goals (including ROI) and schedule for the pilot installation.
- Propose additional public relations and TPM information activities.
- Present draft of master plan.

Appendices
Bibliography and Resources

5S for Operators: 5 Pillars of the Visual Workplace (For Your Organization!) Published by Productivity Press; 1 edition (March 1, 1996). This book is adapted from the work of Hiroyuki Hirano.

Introduction to TPM (another very complete and good article) by J. Venkatesh published by Plant-Maintenance.com at http://www.plant-maintenance.com/articles/tpm_intro.pdf

We would like to acknowledge the groundbreaking work of Nakajima and Suzuki. Much of the information on TPM is derived from the writings of Seiichi Nakajima, Vice Chairman of the Japan Institute of Plant Maintenance. Including: *Introduction to TPM*, and *TPM Development Program*, both by Seiichi Nakajima's (1984 in Japanese and 1988 English translation) published by Productivity Press.

Brady Corporation is a global provider of solutions that identify and protect people, products and premises. The line of lean / visual workplace products includes 5S marking supplies, lean communication boards, as well as software and printing systems that allow users to create their own signs, labels and tags on site and on demand. For more information, go to www.bradyid.com/visualworkplace <http://www.bradyid.com/visual-workplace> or call 1-888-272-3946.

"Just Call Him Mr. Productivity," by John Sheridan, *Industry Week*, May 21, 1990. Mr. Sheridan is the founder of Productivity Press, the major publisher of TPM books.

Lean Thinking: Banish Waste and Create Wealth in Your Corporation by James Womack and Daniel Jones, published by Simon and Schuster, 2003.

Lean TPM: A Blueprint for Change by Dennis McCarthy and Nick Rich, published by Elsevier Ltd., 2004.

Information about the originator of TPM has been excerpted from an article titled "Lessons from the Guru's," published in *Industry Week,* August 6, 1990.

The Machine that Changed the World by James Womack, Daniel Jones and Daniel Roos, published by Rawson Associates, a division of Macmillan Publishing, 1990.

Out of the Crisis by W.E. Deming, published by MIT Press, 1986.

Overall Equipment Effectiveness by Robert Hansen, published by Industrial Press, 2002.

A Study of the Toyota Production System: From an Industrial Engineering Viewpoint (Produce What Is Needed, When It's Needed) by Shigeo Shingo, published by Andrew P. Dillon Productivity Press, 1989.

Total Productive Maintenance, by Terry Wireman, published by Industrial Press, 2004.

TPM: An American Approach, by Terry Wireman, published by Industrial Press Inc, NY, 1991. Mr. Wireman is one of America's leading authorities on TPM. In this book he has adapted Japanese ideas to the American situation.

TPM in Process Industries, Edited by Tokutaro Suzuki, published by Productivity Press 1994

"TPM: More Alphabet Soup or a Useful Plant Improvement Concept?" by William M. Windle and A.T. Kearney, *Plant Engineering*, Feb 4, 1993.

TPM: New Implementation Program in Fabrication and Assembly Industries, Edited by Kunio Shirose, published by Japan Institute of Plant Maintenance, 1996.

"Will the Real TPM Please Stand Up?" by Edward Hartman, American TPM Institute, *Maintenance Technology Magazine,* January 1991.

5S Video
http://www.youtube.com/watch?v=zhnnl8jLTb0
http://www.howcast.com/videos/4783-What-Is-5S
http://www.youtube.com/watch?v=4_p9Yxkn_lM

TPM web site points to many other resources: *http://www.tpmonline.com/*

TPM article library: *http://www.plant-maintenance.com/maintenance_articles_tpm.shtml*

TPM Videos:

http://revver.com/video/1288873/tpm-total-productive-maintenance-lean-manufacturing-tool-ppt/

http://www.youtube.com/watch?v=eI388V_xhws (OEE)

http://www.in.com/videos/watchvideo-tpm-total-productive-maintenance-lean-manufacturing-tool-ppt-2112226.html

http://www.youtube.com/watch?v=o6jo7ENhcFQ (TPM in the office)

http://www.youtube.com/watch?v=zl4aoHrbuoo
(Trip to Japan –Lean Manufacturing)

http://www.youtube.com/watch?v=ZVlKtcJbpvM

Videos on TPS *http://www.tpslean.com/index.htm*

A great archive of single point lessons from Fuss & O'Neill. They are a full-service engineering consulting firm:
http://www.fando.com/News_&_Resources/Single_Point_Lessons/SPL_Archive/

http://www.martsconference.com/ Conference that frequently features TPM speakers.

http://www.kkbooks.com/jishu%20kozen.html Jishu Hozen, Manual Edited by TPM Club India

The web site for the Japanese Institute of Plant Engineering: http://www.jipm.com/

http://wcm.nu/ (World class manufacturing)

Imants BVBA is a Belgium consultancy that is one of the thought leaders in continuous improvement   http://www.managementsupport.com/

Strategos Books & Videos: Chart available of TPM in a Nutshell available at: www.strategosinc.com/training_tpm1-h1.htm

Glossary

Autonomous Maintenance: Routine maintenance and PMs are carried out by operators in independent groups. These groups solve problems without management intervention. The maintenance department is called on to solve bigger problems that require more resources, technology, or downtime.

Asset: May be a machine, a building, or a system. It is the basic unit of maintenance and the driver of the PM and computerized maintenance systems. In this work, the terms *asset*, *machine*, *unit*, and *equipment* are all used interchangeably.

Bolting: The whole field of joint design, fastener position, proper tightening, and bolt choice. For purposes of TPM, bolting usually refers to tightening and inspection of bolted joints for tightness.

Breakdown maintenance: See *emergency work*

Call back: A job for which a maintenance person is called back because the asset broke again or the job wasn't finished the first time. See *rework*.

CI (Continuous improvement): Small improvements that over a period of time compound into a major stream of waste reductions. Continuous mean ongoing — we never stop CI.

CMMS (Computer Maintenance Management System): System used to store maintenance information, issue work orders, and manage PMs. It is the business system for maintenance.

Continuous Improvement (in maintenance): Reduction of the inputs (hours, materials, management time) to maintenance to provide a given level of service. Can also mean increases in the number of assets, or use of assets with fixed or decreasing inputs.

Corrective maintenance (CM): Maintenance activity that restores an asset to a preserved condition. CM is normally initiated as a result of a scheduled PM or PdM inspection. See also *planned work*.

Deferred maintenance: All the work you know needs to be done that you choose not to do. You put it off, usually in hopes of retiring the asset or getting authorization to do a major job that will include the deferred items. You worry that the asset will fail before you get to it.

DIN work: 'Do It Now' is non-emergency work that you have to do now. An example is moving furniture in the executive wing.

Emergency work or emergent work: Maintenance work requiring immediate response from the maintenance staff. Emergent work also refers to work that emerges after you open up an asset (pump, vessel, etc). Emergency work is usually associated with some kind of danger, safety, damage, downtime, or major production problems. Machine or asset does not provide the function it was designed for. Usually due to a part failing, a design flaw, or misuse; also called *reactive maintenance.*

Failure modes: All the ways an asset can lose function. Not all failure modes are maintenance related.

GLO (Generalized Learning Objective): The general items necessary to know to be successful in a job. Each job description is made up of a series of GLOs.

Gemba: A Japanese term meaning "the actual place" or "the real place." In business, *gemba* refers to the place where value is created; in manufacturing the *gemba* is the factory floor.

Genbutsu: The part itself, the actual thing you make.

Genchi Genbutsu: This is one of the drivers for all of TPM; it means "go see for yourself." The idea is that any reporting will be a simplification so that any solutions done remotely might be ill advised. TPM is hands on so there is no translation for problem solving.

Gensho: The phenomenon.

Iatrogenic Formal definition: Induced inadvertently in a patient by a physician's activity, manner, or therapy. In our case, it describes failures that are caused by your service person.

Inspectors: The special crew or special role that has primary responsibility for PMs and PdMs. Inspectors can be members of the maintenance department or of any related department (machine operators, calibration, drivers, security officers, custodians, etc.).

JIT (Just in Time): An inventory strategy that strives to reduce in-process inventory. To meet JIT objectives, the process relies on signals that tell production when to make the next part. JIT can dramatically improve a manufacturing organization's return on investment, quality, and efficiency (part of TPS).

Jishu Hozen: Japanese phrase for *Autonomous Maintenance.*
Kaizen: Meaning "improvement" in Japanese. Kaizen is used to refer to practices focusing on continuous improvement. A focused kaizen that is designed to address a particular issue over the course of a week is referred to as a "kaizen blitz" or "kaizen event."

Lean Maintenance: "An all out war against waste in maintenance" Shigeo Shingo (and part of TPS).

Life Cycle: Denotes the stage in the life of the asset. Three life cycles or stages are recognized by the author: start-up, wealth, and breakdown.

Management: The act of controlling or coping with any eventuality.

Maintainability Improvement: Also *Maintenance Improvement*. This activity looks at the root cause(s) of breakdowns and maintenance problems, then designs a repair that prevents breakdowns in the future. It includes improvements to make the equipment easier to maintain.

Maintenance: Although the dictionary definition is "the act of holding or keeping in a preserved state," it doesn't say anything about repairs. It presumes that we are acting in such a way as to avoid the failure by preserving the asset.

MP (Maintenance Prevention): Redesigning a machine so that it needs less maintenance and the source of the maintenance problem is repaired.

Muda: An activity that is wasteful and doesn't add value or is unproductive. One of the three wastes in TPS.

Mura: term for unevenness, inconsistency in physical matter, variations in product or output. One of the 3 wastes in TPS

Muri: Unreasonable, a Japanese term for overburden, unreasonableness, or absurdity, that has become popularized by its use as a key concept (along with Mura and Muda above) in the Toyota Production System. Muri can be avoided through *standardized work.*

OEE (Overall Equipment Effectiveness): A ratio of what an individual machine is putting out divided by the individual machine's theoretical ideal output. All losses are catalogued.

OPE (Overall Plant Effectiveness): A ratio of what the plant is putting out divided by the plant's theoretical ideal output. All losses are catalogued.

Parts: All the supplies, machine parts, and materials needed to repair an asset, or a system in or around an asset. In some cases, parts are separated from supplies and consumables.

Pareto: Italian 19th century economist whose work concerned the leading families in Italy controlling a bulk of the wealth. We've turned his work into the 80/20 rule where 80% of the action is related to only 20% of the population.

PM (Preventive Maintenance): A series of tasks that either extend the life of an asset or detect that an asset is at a critical point and is going to fail or break down.

PM frequency: How often the PM task list will be done. The frequency is driven by the PM clock. See *frequency of inspection.*

PMO (PM Optimization): A process of matching the PM task with the failure mode and selecting the best person to do the task.

Predictive Maintenance (PdM): Maintenance techniques that inspect an asset to predict when a failure will occur. For example, an infrared survey of an electrical distribution system might look for hot spots (which would be likely to fail). In industry, predictive maintenance is usually associated with advanced technology such as infrared or vibration analysis.

Proactive: Action before a stimulus (the antonym is *reactive*). A proactive maintenance department takes actions before a breakdown occurs.

RCM (Reliability-centered Maintenance): A maintenance strategy designed to uncover the causes and consequences of breakdowns. RCM sets up priorities by the severity of the consequences. PM tasks and redesign are directed specifically at those failure modes that have the worst consequences. RCM is a procedure for uncovering and overcoming important failures.
Reactive maintenance: see *breakdown maintenance*

Rework: All work that has to be done over. Rework is bad and indicates a problem with materials, skills, or scope of the original job. See also *call back.*

Root cause (and root cause analysis — RCA): The underlying cause of a problem. For example, you can snake out an old cast or galvanized sewer line every month and never be confident that it will stay open. The root cause is the hardened buildup inside the pipes. Analysis would study the slow drainage problem and figure out what was wrong. The study would also estimate the cost of leaving the defective pipe in place. Some problems (not usually this type of example) should not be fixed.

Routine work: Work that is done on a routine basis where the work and material content are well known and understood. An example is daily line start-ups.

SLO (Specific Learning Objective): The detailed knowledge, skill, or attitude necessary to be able to do a job.

Seiketsu: Standardization (the 4th S of 5S).

Seiri: Separating and removing all items not needed (the 1st S of 5S).

Seiso: Shine the workplace — clean everything up (the 3rd S of 5S).

Seiton: Organize "Each item has a place, and only one place" (the 2nd S of 5S).

Shitsuke: Self discipline (the 5th S of 5S).

Short Repairs: Repairs that a PM or route person can do in a short time with the tools and materials that they carry. A short repair is a complete repair that can be done in a short time. It is different from a temporary repair.

SOP (Standard Operating Procedures): A listing of the step-by-step instructions to perform a process or use a machine.

TLC (Tighten, lubricate, clean): Basic maintenance practices of TPM.

TPM (Total Productive Maintenance): TPM is a maintenance system set up to eliminate all of the barriers to and losses in production. TPM identifies production losses and uses production operator teams to solve the problems causing the waste. Autonomous maintenance teams (focusing on operators) are used to carry out most basic maintenance activity.

TPS (Toyota Production System): A management system designed to eliminate waste and promote flexible production pioneered at Toyota automobile company.

Task: One line on a task list (see below) that gives the inspector specific instructions to do one thing.

Task List: Contains specific directions to the inspector about what to look for during that inspection. Typical tasks include inspect, clean, tighten, adjust, lubricate, and replace.

Work Order: Written authorization to proceed with a repair or other activity to preserve an asset.

Work request: Formal request to have work done. Work requests are generally filled out by a maintenance user. Work requests are usually time/date stamped and form the basis of the work orders.

Examples of In-Depth Training

Lubrication

Lubrication is essential for long life in equipment. The field has advanced a long way from the old grease gun days. As an example of the complexity of the field, Exhibit A-1 shows the curriculum from Petroleum Management Consultants (http://www.sandybrae.com/training/index.htm) for a 5-day class in hands-on lubrication:

Exhibit A-1   Curriculum for Course on Lubrication

Course Introduction How Important is Price Viscosity-Temperature Charts	Additives Definitions Additive Exercise Snake Oil by Fred Rau AW and EP Exercise
Petroleum Chemistry Chemistry Fundamentals Refining Process Refining Terms Refining Slide Presentation Industry Data Annual Sales of Lubricants Base Stocks	Benefit Selling (For Sales Personnel Only) The Product Sell Benefit Selling Proposal Writing Writing Workshop Ten Principles of Clear Writing Report Techniques
Grease NLGI Film on Grease Lab Exercise on making Grease Lubricant Selection General Trends Trabon Centralized Grease System Greased Bucket Pins Furnace Reheat Door	Oil Analysis - Are you wasting money? Oil Analysis Terms Reasons for Oil Analysis Program Failures On-Site Oil Analysis Program Filter Patch Interpretation Guide Slide Show Laboratory Oil Analysis for Off-Highway Equipment Oil Analysis Interpretation Tips Engine Oil Analysis Interpretation Guide Diesel Engine Oil Analysis Check List Donaldson Oil Analysis General Guidelines Oil Analysis Sheets
Synthetics Types and Properties Applications Comparative Temperature Limits Synthetic Engine Oil Exercise Synthetic Cost Savings Engine Oil Drain Savings Cost Comparisons	Turbines Slide Show Steam Turbine Slide Show Insert Gas Turbine Slide Show Insert Trouble Shooting Problems in EHC Control Fluid Hydraulic Systems

continued on next page

Fundamentals of Lubrication
Friction
Lubricant Film Definitions
Thick Film Lubrication
Thin Film Lubrication
Solid Film
Lubricant Films

Machine Elements
Bearings
Cement Mill Grease Exercise
Bearing Identification Exercise
Bearing Review
Bearing Types Defined
Bearing Characteristics
Oil Viscosity Lubrication Selection
Grease Re-Lubrication Period
Relubrication Guidelines for Bearings
How much grease is enough?
Viscosity Selection for Rolling Element Bearings
Oil Viscosity Selection for Plain Bearings
Gears
Gear Maintenance Concepts
Air Padding
Removing Wear Metals
Gear Tooth Terminology
Marion Hoist Gear
Steel Mill Lubricant Recommendation Exercise
Gear Identification
Lubricant Considerations for Gearboxes
Gearboxes
AGMA Lubricant Designations
Sae Gear Oil Classifications
Misc. Machine Elements
Chains, Coupling, Wire Rope, Pistons and Other
     Components Slide Show

Internal Combustion Engine
Internal Combustion Engine Parts Slide Show
Lab Exercise with Internal Combustion Engine
Lab Exercise with Engine Parts
Engine Oil Consumption Concepts
Engine Clearances and Tolerances Slide Show
Engine Exercises

Developing Benefit Areas
Cost of Leaks
Mobile Equipment - Typical Data
Mobile Equipment - Typical Data
Contractor Fuel and Lubricant Requirements
Construction and Mining Industry Overview
Earth Moving Lubricant Consumption Guidelines
Strip Mining Lubricant Consumption Guidelines
     for Eastern Coal
Crushed Stone Industry Cost Data
Surface Mining Requirements (Lubricants)
Underground Mining Requirements (Lubricants)
Exercises
Justification of Synthetic Lubricants in Fleets

Power Train
Chassis Components Slide Show
Lubricants for Non-Driven Axles
Trouble Shooting
Recognizing Cost Reduction Opportunities
Machine Tools and their Operations

Hydraulics
Hydraulics Slide Show
Hydraulic System Basics
Guidelines for Vane Pumps
Temperature/Viscosity Exercise
Article
How to detect and prevent hydraulic pump cavitation
Lab Hydraulic System Demonstration
Lab Hydraulic Pump Exercises
Converting old ACFTD particle sizes to NIST Sizes
ACFTD vs. NIST Calibration
Beta Ratios Comparing ACFTD vs. NIST vs. MTD Multi-Pass
Clearing Up Hydraulic Fluid Myths
Filtration Concepts
ISO Particle Count Classification
Typical Hydraulic Component Cleanliness Requirements
Napa Micron Rating for Fluid Filters
Vickers Target Cleanliness Ratings
Compressors
Introduction
Slide Show Insert
Discussion Issues
Air Compressor Analysis (Trouble Shooting)
Reciprocating Air Compressor Operating Data
Rotary Type Compressor Operating Plant Hydraulic
     Study Outline

Forms
Storage and Handling
Lubrication Survey Form
Abbreviations used in Lube Surveys
Plant Hydraulic Study
Hydraulic System Inspection Report
Hydraulic Fluid Inspection
Hydraulic System Inspection Checklist
Mobile Equipment Inspection Form

Dragline Hoist Gear Exercise
Heavy Equipment Cost Benefit Areas
Developing Cost Saving Benefits for Industry
General - What are Suffering Points
Finding Suffering Points
Air Compressors
Air Intake Systems (Compressors and Diesel Engines)
Conveyor Systems
Cement or Ready Mix Trucks
Electric Motors
Forklift Truck Maintenance
Gearboxes
Hydraulic Systems
Lubricant Storage and Handling
Storage and Handling Slide Show
Savings on Lubricant Storage and Handling with Bulk Oil

One might argue that this level of training is overkill for an operator, and the U.S. $1800 tuition is a lot of money. On the other hand, significant numbers of breakdowns come from ignorance of the basics in one of the key TPM areas of concentration. I'll bet there would be benefit from your fitters and millwrights to take this type of class too.

Another Important Area of Training: Bolting

This example is included to impress upon TPM aspirants the amount of information needed to master a seemingly simple field like bolting. It would be a great idea for the maintenance department to take these kinds of classes too! Bolt Science (www.bolting.info) offers a complete training in Bolting, as summarized in Exhibit A-2.

Exhibit A-2 Sample Curriculum in Bolting

Introduction to Threaded Fasteners
Know the meaning of thread terminology.
Learn when it is appropriate to use a fine rather than a coarse thread.
Be aware of the principal bolt and nut strength property classes and
how they should be specified.
Learn to match the nut strength to that of the bolt so that thread
stripping problems are prevented.
Find out why bolt tensile fracture is preferable to the threads stripping.
Learn what the proof load is and why it is used.
Be able to identify the meaning of the markings on bolt heads and nuts.
Learn about the thread stress area and how it is derived and used.
Be able to calculate the tensile strength of a threaded fastener.
Understand how a pre-tensioned bolted joint sustains an applied load.

Preload Variation in Threaded Fasteners
Learn why there can be such a significant variation in the preload (tension in the bolt) and the consequences of this.
How the torque is distributed between the threads and the nut face when free spinning and torque prevailing fasteners are used.
Why preload is so crucial in a bolted joint.
How preload variation can be accounted for at the design stage.
Understand the effect of the tightening method on the preload variation sustained by a fastener.

Galling of Threaded Fasteners
What is galling and what types of materials tend to be affected?
Examples of fastener threads that have galled, sectioned, and x-ray photos.

Four Ways that Galling Can Be Eliminated

Methods of Tightening Threaded Fasteners
Have an understanding of the principles behind each of the following tightening methods:
Torque controlled tightening
Torque-angle controlled tightening
Yield controlled tightening
Bolt stretch method
Heat tightening
The use of load indicating methods
The use of ultrasonics to determine bolt loading

Failure Modes of Threaded Fasteners
Learn the differences between a manufacturing and design quality defect.
Be able to identify whether a failure is due to a fault in the design specification or is manufacturing related.
Learn the five main design-related failure modes of threaded fasteners and bolted joints.
Have knowledge of the critical importance of a fastener's clamp force in ensuring a joint's structural integrity.
Understand why the joint design normally prevents bolt overloading.
Learn about fatigue and where failures normally occur on a threaded fastener.
Understand why bearing stress can be crucial in ensuring a reliable joint.
Learn about the nature of internal and external thread stripping failures.

Vibration Loosening of Threaded Fasteners

Have an overview of the research completed over the last 50 years into
establishing the cause of self-loosening of threaded fasteners.
Appreciate the forces that are acting on the threads that tend to self loosen a fastener.
Explain why fine threads can resist loosening better than coarse threads.
State the inclined plane analogy.
Learn about the work completed by Goodier and Sweeney into loosening due to variable axial loading.
Describe the work completed by ESNA and the theory of shock induced loosening and
 resonance within fasteners.
Understand the MIL-STD 1312-7 vibration test for fasteners.
Junker's theory on self-loosening of fasteners and why fasteners self-loosen.
The Junkers/transverse vibration test for fasteners.
The influence that vibration amplitude has on the fastener self-loosening rate.
Preload decay curves and the effectiveness of various fastener types in resisting vibrational loosening.
The findings of Haviland and Kerley and how fasteners can come loose as a result of bending, shock or impact
and differential thermal expansion.
Conclusions from the research and how loosening can be prevented.

Torque Control
What is meant by a tightening torque?
Units used to measure torque.
What are the consequences of not applying sufficient torque to a bolt.
How torque is absorbed by a nut/bolt assembly.

The torque-tension graph.

The relationship between the tightening torque and the resulting bolt preload (tension).

The factors which affect the torque-tension relationship.

The nut factor method of determining the correct tightening torque.

Example calculation of how to determine the correct tightening torque.

Scatter in the bolt preload resulting from friction variations.

Determining the bolt preload (tension) resulting from a tightening torque.

Prevailing torque fasteners (such as those containing a nylon insert) and how they affect the torque distribution and the correct torque to use.

Load Sensing Fasteners

The use of strain gauged bolts.

The use of load cells.

The use of Rotabolts™.

The use of Smartbolts™.

The use of direct tension indicators (load indicating washers).

Squirter™ direct tension indicators.

Tension control bolts.

Hydraulic Tensioning of Threaded Fasteners

The principles behind hydraulic tensioning.

The number of tensioners that are used to tighten a joint – 100%, 50%, 33% and 25% tensioning methods.

The effect of elastic recovery on the tension induced into a bolt.

The use of hydraulic nuts and the sequence used to tighten them.

The use of oil filled nuts.

The use of rubber filled nuts.

Tightening Procedures

The problems of tightening multi-bolt assemblies.

Elastic interaction or bolt cross-talk.

The use of a tightening sequence.

The single pass tightening sequence.

Tightening sequences for non-circular bolt patterns.

Tests completed to verify tightening sequences.

The two pass tightening sequence.

The use of multiple tightening tools.

Bolt cross talk and hydraulic tensioning.

Methods that can be used to check the tightening sequence.

The solder plug method.

The use of pressure sensitive films.

Establishing a tightening procedure.

Examples of tightening sequences for circular joints consisting from 4 to 32 bolts are given in the handbook together with an example tightening procedure.

Exhibit A-3 Partial List of Seminars
from Joel Levitt

Joel Levitt

E-Mail: **jdl@maintrainer.com**

Visit us at **www.maintrainer.com**.

Sign up for our free Maintenance Manager's Newsletter.

International: 610-397-1006   U.S. Toll free: 1-800-242-5656

Title	Length	Short Description
Lean Maintenance	½, 1, or 3 days	This workshop actually introduces cost reduction and carries it out. This course is for maintenance craftspeople or to train Lean trainers and facilitators.
Maintenance Management	2 days	Introduction to maintenance management concepts including benchmarking, TPM, economic analysis, PM, RCM, planning. Can be designed with additional topics
PPM (Preventive and Predictive Maintenance)	1, 2, or 5 days	Complete discussion of all aspects about PM. Longer courses add depth and the 3-day adds case studies from your plant.
Maintenance Planning	1, 2, or 5 Days	Complete course and hands-on workshop on planning. Includes all steps in planning work for a maintenance department
Maintenance Shutdown and Turnaround Planning and Management	1, 2, or 5 days	Complete course and hands-on workshop in planning and managing large maintenance projects and shutdowns.
TPM (Total Productive Maintenance)	1 or 3 days	Introduction to TPM in a factory environment.
Optimizing the Maintenance Inventory	2 days	Review of all the issues surrounding running an effective inventory including theory, economics, EOQ, capital spares, rebuildables.
Management Skills for Maintenance Supervisors and Team Leaders	2 or 3 days	Topics include: supervision style, delegation, planning, PM, PPM, basic time management.
Advanced Management Skills for Maintenance	2 or 3 days	Topics include: advanced time management, communications, craft training. Can be run with above for 4, 5, or 6 days
Maintenance Management for Buildings and Facilities	2, 3, or 5 days	Complete course to manage maintenance building complexes (airports), venues (stadiums), office buildings, etc. Does not duplicate Maintenance Management.
Maintenance Management for Factories and Process Industry	2 days	Complete course to manage factories, refineries, batch plants. Does not duplicate Maintenance Management.
Breakthroughs in Fleet Maintenance	2 or 5 days	Complete course to manage truck, car, train maintenance. Does not duplicate Maintenance Management.
World Class Maintenance	2 hours	Talk on the attributes of world-class maintenance. Used by organizations to inspire the maintenance people
Time Management for Maintenance Professionals	2 hours	Training on using the little time available to good effect. Includes 4 projects to improve time management and new thought models to improve achievement.
Specialized speeches, workshops, or courses designed on request	1 hour to 10 days	Contact office for conversation and a quotation.

Index